TEXTURE

TEXTURE

HUMAN EXPRESSION IN THE AGE OF COMMUNICATIONS OVERLOAD

RICHARD H. R. HARPER

THE MIT PRESS
CAMBRIDGE, MASSACHUSETTS
LONDON, ENGLAND

For information about special quantity discounts, please email special_sales@mitpress.mit.edu.

This book was set in Engravers Gothic and Bembo by Toppan Best-set Premedia Limited. Printed and bound in the United States of America.

Library of Congress Cataloging-in-Publication Data

Harper, Richard, 1960–
Texture : human expression in the age of communications overload / Richard H. R. Harper.
p. cm.
Includes bibliographical references and index.
ISBN 978-0-262-08374-4 (hardcover : alk. paper)
1. Communication—Technological innovations—Social aspects. 2. Communication—Social aspects. 3. Personal communication service systems—Social aspects. I. Title.
HM1166.H37 2010
303.48'33—dc22

2010003176

10 9 8 7 6 5 4 3 2 1

CONTENTS

ACKNOWLEDGMENTS vii

1 INTRODUCTION 1

2 THE COMMUNICATIONS PARADOX 9

3 ABSENCE TO PRESENCE 59

4 PARADOXICAL DELIGHTS 109

5 SOMETHING TO TELL 153

6 THE SEAMS THAT BIND 193

7 THE TEXTURE OF AN EXPRESSIVE FUTURE 229

INDEX 273

ACKNOWLEDGMENTS

Three people helped enormously in the writing of this book. All three I have had the pleasure of meeting through work—Lynne Hamill, at the Digital World Research Centre in Surrey; Dave Randall, at Lancaster University; and Rod Watson, at Manchester University. Each offered rather different kinds of assistance: Lynne insisted on the facts, Dave insisted that I get it right, and Rod made me read properly and more comprehensively. Even so, they would all deny any responsibility for the book—and quite rightly. If they had had their way, it would have been better. Alas, what is here is the best I could do. Nevertheless, it is for them.

All books derive from the tender tolerance of friends and colleagues for the vain obsession of the author. This book is no exception. As its author, my debts are varied and great. Beyond those to whom it is dedicated, a special thanks go to Leopoldina Fortunati and Ronald Rice for their extensive and detailed comments on the manuscript that led to major improvements; to Gary Marsden and Matt Jones for more general comments; to James Katz, for introducing me to John Durham Peters's *Speaking into the Air* (Chicago, 1999) and Ernest Becker's

The Birth and Death of Meaning (Glencoe, 1962), both of which altered my thinking on the topic of the book. Others provided various bits and pieces that were invaluable. Phil Fawcett, for example, provided data about Microsoft email volumes; Paul Dourish, information about some early applications used in Xerox EuroPARC; and Dave Kirk, some grumpy comments about the definition of *ergonomics*.

Some of the materials presented in the book were derived from research studies that I conducted with colleagues. Of these, the most important debt is owed to three people: Alex Taylor, who worked with me on studies of mobile phones (discussed in chapter 4), Stuart Taylor, who built the Glance-phones (discussed in chapter 5), and Phil Gosset, who has for many years worked with me on various aspects of the social shaping of mobile telephony. His influence shows itself throughout the book. Other work that has suffused the book has entailed collaborations with Sian Lindley, Tim Regan, Shahram Izadi, Laurel Swann, Simon Rubens, Mark Rouncefield, Wes Sharrock, and Jane Vincent.

Beyond this, there is also a debt of gratitude due to all my colleagues at Socio-Digital Systems (SDS) at Microsoft Research, Cambridge. They have had to put up with me being an absent member while the scribbling for this book was done. Of those not previously mentioned, I would like to note Richard Banks especially and, outside of SDS, many colleagues and friends in the lab. Thanks all.

Lastly, I must share my gratitude to my closest colleague of all—Abi Sellen. She hasn't helped in the writing of this book, and I made a special effort not to turn to her for help because she helped in the writing of many other books in the past. But without Abi and her constant presence, the world I inhabit would be much smaller—indeed, would lack what makes it worth being in.

Sections of chapter 1 appear in "Absence to Presence: A Vision of the Communicating Human in Computer-Mediated Communications Technology Research," in J. Höflich, G. F. Kircher, C. Linke, and I. Schlote, eds., *Mobile Communication and the Change of Everyday Life* (41–57) (Berlin: Peter Lang, 2010). Sections of chapter 3 appear in "From Tele *Presence* to Human *Absence*: The Pragmatic Construction of the Human in Communications Systems Research," in *Twenty-third Annual Conference of the British HCI Group (HCI 2009)* (London: British Computer Society, 2009), copyright the author. Parts of chapter 4 appear in A. Taylor and R. Harper, "The Gift of the Gab: A Design-Oriented Sociology of Young People's Use of Mobiles," *CSCW: An International Journal* (2003): 267–296, copyright Elsevier. Parts of chapter 3 appear in "Are Mobiles Good for Society?," in K. Nyiri, ed., *Mobile Democracy: Essays on Society, Self and Politics* (185–214) (Vienna: Passagen Verlag, 2003), permission from Professor Nyiri. Parts of chapter 5 appear in "Glancephone: An Exploration of Human Expression," *Eleventh International Conference on Human-Computer Interaction with Mobile Devices and Services (Mobile HCI 2009)* (New York: ACM Press, 2009), copyright the authors. Much of the evidence in chapter 5 derives from a study by my colleagues published as A. Sellen, A. S. Taylor, J. Kaye, B. Brown, and S. Izadi, "Supporting Family Awareness with the Whereabouts Clock," in P. Markopoulos, B. de Ruyter, and W. Mackay, eds., *Awareness Systems: Advances in Theory, Method and Design* (225–242) (London: Springer-Verlag, 2007).

TEXTURE

1 INTRODUCTION

PREAMBLE

In my own company, Microsoft, each employee sends and receives about 120 emails every day. Many also receive alerts from really simple syndication (RSS) feeds; and most run Messenger, our own instant messaging client. At Microsoft, we like to think that we are busy, efficient, effective, and knowledgeable enough about the communications technologies of the twenty-first century to leverage them for our own benefit. After all, we like to think that we helped invent some of them, and if not, we certainly have a business interest in most. We should know about these things. Yet my colleagues complain that they are constantly interrupted, that they can't keep up with their emails, that they find it difficult to say goodbye when instant messaging, and that they don't have enough time to get their work done. Somehow the balance of things seems to have gone wrong, they explain to me. The tools designed to let them work better seem to have had the opposite effect. And it is not only at work that this malaise seems to be appearing. My colleagues say that when they leave work, their personal cell phones start bleeping as text messages arrive.

"There are voice messages, too," they complain. Worse, when they get home, there are traditional letters—not many, to be sure, but always some—and these also have to be dealt with. So they say to me that at work there is no time for work and at home there is no time for being at home. The point of their complaints is that their world (which is my world and the world that most readers of this book occupy) seems to be getting harder to live in. It is busier than ever and fraught with more things said and communicated than ever before. It is no surprise, then, that each morning over coffee my colleagues assert to me, "Surely a threshold is being reached. Enough, already. No more communication!"

In corporate and academic settings, this issue—the idea that some kind of tipping point beyond which the balance between what is practical and what is excessive has been or is about to be reached—is well known. The phrase *communications overload* is commonly heard. So it is hardly surprising that many researchers are devising tools and techniques that can reduce this problem. Various kinds of solutions have been proposed. Some researchers are devising machine learning applications that assess whether a change in the content of a Web site is sufficiently interesting to alert (via RSS feeds) a user. Others are devising filtering mechanisms that can let users triage their in-boxes more effectively. Yet others are designing ways of integrating messaging channels to reduce the burden of dealing with them all. Some of these solutions are, by their authors' own admission, forms of fire fighting. Assessing the degree of change in an RSS feed seems to be a case in point. All this does is put off to the future the moment when a user says, "That's it. No more feeds." Similarly, new ways of filtering and triaging only delay the day when the limits of time press down. One can hear future users grumbling: "When does one deal with the less urgent if all one ever has time for is the urgent?" and "What about the simply important if not urgent?" Other ideas have more merit, even if they are evidently not permanent solutions. Allowing people to easily access their messages at any time and place comes to mind. This certainly is one of the appealing properties of BlackBerrys (although making access 24/7 might increase email volume).

Despite all their protestations about communications overload, many researchers who are undertaking projects to find these and other solutions are doing something else that, at first glance, seems perplexing. For one thing, these attempts at solving the communications overload are not the primary focus of their research endeavors. Indeed, between their continuous emailing and instant messaging, these researchers spend much of their time adopting new ways as they arise. As I write this, for example, I find my colleagues keeping their newly acquired Facebook accounts up to date or creating short messages via Twitter on their cell phones. And as their goal at work, they devise new ways of communicating. They seek ways of conveying tactile experiences, for example, as supplements for the audiovisual messaging that they have spent much time refining over the years. They devise new social communications systems that let people vote, comment, and express to large groups.

Their delight in seeking new ways of expression is reflected in their home lives, too. When asked, they do not say that they are fighting off the torrent of messages and communications that they claim are terribly irksome. Instead, they have been writing and reading blogs and uploading files to YouTube. They also say that they have been emailing friends from university, for instance, with whom they had lost contact but whose details they found on a Web site. In other words, they have not been seeking peace and quiet and the solicitude of private reflection but have been adding to the volumes of communication. They have been actively creating that content and seeking new ways of producing it. In other words, they have been delighting in the very thing that they seem to complain about: they gleefully produce the content that at other times they say weighs them down. At work and at play, they fill up their lives with the thing that they say stops them from working and playing. They communicate yet complain about communication. They express themselves in new ways yet berate the fact that there is not enough time to listen to others' expressions.

This doing of one thing and saying of another might seem to be an amusing albeit lamentable fact of modern lives. The evocation of this world, as illustrated with my own workplace, could be merely a flavorsome way of conveying what it means to be someone engaged in contemporary existence, professionally and personally. Life is busy, but people are getting on with their lives, leveraging what they can to be in touch and keep in touch in the frenetic, networked world of the twenty-first century. We are all too busy these days, but what more can one usefully say?

I think one can say something about where we have come from, how we got here, and where we might go in the future. I think that what needs to be said has to do with our desire to communicate and express, a relationship between this and our ability to devise and exploit new technologies that foster and enable that same expression, and a philosophy about what it means to be human in this day and age. The purpose of this book is to explore these issues. It seems to me that there is a conundrum to be explored that has to do with the tension between communications overload and the desire for communication, between the boredom that older technologies of communication induce and the fascination that exploring the properties of new ones cultivates, and the possibility that communication imposes on us a need to respond, to act, to answer the expressions of others. This tension lies at the heart of our current circumstances. Analysis of how we got here, how technological innovations and socially creative ways of exploiting technology interact, how the delight experienced in using new expressive forms of communication is counteracted by the vexation that the resulting moral burdens place on us—all of this can provide the basis for insights into what it means to be alive, connected, and expressive and into where we might want to be at the end of the next decade and in the years thereafter.

The conundrum at the heart of this book is not simply whether we have reached a point of communications overload. We have worked hard over the past century or so to create different ways of communicating and expressing ourselves, and we have wrapped ourselves up in a social universe of communicative obligation. In seeking new forms of expression, we don't seek an end point where we have expressed all that there is to express. Rather, we define new acts of communication that lead others in various ways and with various consequences to their own acts and their own responses. To communicate is to foster communication.

These ideas did not necessarily drive the inventiveness that helped create the current landscape. What motivated technological inventiveness, was (and still is for some, no doubt) partly a presumption that more, richer, or quicker communications technologies will replace older, slower, and less rich means of expression. It is also partly a result of the inventions in question (some of them, anyway) being appropriated by people desiring (and fostering a desire) to indulge in and exploit new channels of communication. The desire here (one that in its satiating creates greater intensity) is not one that seeks to replace older technologies or to make communications more efficient as if the aim of this desire were to make the human machine optimal. The desire is for supplementing and enriching the expressive vocabulary of human experience.

The result of these two countervailing tendencies—one to create substituting technologies and the other to delight in diversity and an aversion to substitution—is that we now live in a world where there is a *texture* to our communicative practices that is manifest in the different ways we experience and exploit our communications technologies. We choose one means of communication over another because the expression that it enables is taut and quick and brings those we communicate to via that channel closer to us in that order—tautly and quickly. We choose another because it is loose and slow—gentle—and so treats those we express to gently in turn. We select a third because it is permanent and inviolate: however much those we are communicating to try to avoid that missive, they will find it cannot be undone. We choose a fourth because it is ephemeral, although we offer it as a token of regard. Our ways of expressing are strategic and are binding us together in different ways. As we move forward and orient to ways of creating a different future and a different texture, we must recognize that there is no end point to such endeavors. We would do better if we saw that the future we are creating is one that is more social and in which human expression is becoming ever more central to what we are about and how we understand ourselves. We need to recognize too that this self-regard might—and indeed, probably does—come at the cost of reducing other forms of human action.

I do not suggest, however, that this is an analog of Jean Baudrillard's dystopia, in which the media corrupt the essence of things as they are (as he argues, for instance, in his 1970 book, *The Consumer Society: Myths and Structures*). Nor do I suggest that the new communications-rich landscape should lead us to reconsider or redefine what we understand as the thing doing the communicating—the human. This is certainly the view that some commentators take, such as Kenneth Gergen in his book *Relational Being: Beyond Self and Community* (2009). His view is that we have constructed our society on an eighteenth-century premise that there is an essence—a fixed concrete thing that is us, ourselves. In his view, this has diminished the potential value of dialog and the importance of interaction and communication in human praxis (human doings, in more prosaic words). Although he does not claim that the technologies of our current time are forcing a reconsideration of this premise, he does say that our practices are encouraging us to reconsider whether this starting premise is helpful, even leading us to ask whether it might be wrong. He suggests that we might renew our social relations if we made central our conversational or dialogic identity, not our inner, fixed soul. Again, as with the consumer dystopia view, I will argue that our experience of communications richness and variation does not lead us to redefine the metaphysics of human nature in just this way, although it ought to lead us to reflect on what our communicative performances achieve.

This book allows me to comment on these concerns without coming to a fixed judgment. My reflections are designed to enable readers to form their own views. My purpose is to explain how the technologies of communication are the means whereby we invigorate, shape, and alter the very experiences of what it is to be human. In this view, we are, to some degree, *what we say*, but we change as we invent new ways of *doing* the saying. Whether we ought to condemn our choices (as the consumer theorists imply) or whether we should reconfigure our sense of self (as the social psychologists say) is not a question that I answer directly—although I hope to offer reasons and evidence about what the answers (if answers are to be found) might require. It is for readers to make judgments. If my thesis is right, then they are casting themselves in new light as they express. It is for them to ask who they want to be and, beyond this, what kind of society they want to help create.

REFERENCES

Baudrillard, J. 1970. *The Consumer Society: Myths and Structures*. Trans. C. Turner. London: Sage.

Gergen, K. 2009. *Relational Being: Beyond Self and Community*. Oxford: Oxford University Press.

2 THE COMMUNICATIONS PARADOX

PREAMBLE

When my mother admonished me to write home as I was leaving for university, I had no idea that the moral implications of this phrase would still be resonating twenty-five years later. I left home full of enthusiasm and certainty, not philosophical doubt. Nor did I think that communication would be a concern of my professional life. The word itself is a veritable catchall for all sorts of acts and forms of life. Computers communicate to each other, and so do snails. Poets communicate to other poets, and cars traveling through an automated toll booth communicate to a toll meter. They are all passing information. They are all communicating. But the fact that we can use the same word to describe these actions is misleading. What each entails is different in all these examples. One of the most celebrated philosophical rows of the 1970s took place precisely over the meaning of the word. As Jacques Derrida, the French literary theorist, asked in the first line of his essay "Signature Event Contest" (1972, 1), "Is it certain that to the word *communication* corresponds a concept that is unique, univocal, rigorously controllable, and transmittable: in a word, communicable?" His concern was to make a mockery of any attempt to say that it did, and his target was speech-act theory, derived from the ordinary-language philosophy of John L. Austin (see *How to Do Things with Words*, 1962). Without getting involved in any similar sort of deconstruction, I acknowledge that human communication is a slippery topic and has become all the more so in the age of communications overload. I started this book by asserting that each of my colleagues at Microsoft receives 120 emails per day, which was accurate when I started this book about a year ago (since then, the number has increased substantially, I am told). That figure reflects the millions of emails that Microsoft receives (according to our computer postmaster in Redmond, Washington) divided by the number of employees (95,000). The only way to make such a measure is to count emails in and emails out at the junction or interface between my organization and the outside world, not to ask all 95,000 of my colleagues. This means that some people receive more than 120 emails and some fewer. My ultimate boss's email address name, *BillG*, receives many times more emails than anyone else. When people want to complain about Microsoft's products, rail against the cruelties of capitalism, or ask for a charitable contribution, they send an email to Bill Gates. He gets considerably more than 120 a day, just as most days I get considerably fewer.

Even within these obvious differences, other factors occlude an accurate measure. Huge numbers of emails are junk (junk mail), and at Microsoft we have filters that stop most (indeed, millions) of these. But filters filter out the good and the bad, which means that some emails sent from a real person with a genuine need are trapped and filtered. And individual users can set their own filters to remove certain emails from their Outlook accounts and Exchange servers. The figures that I used to calculate the per capita distribution of emails to my colleagues did not include the emails removed by corporate or personal filters. According to some estimates, junk emails are a major proportion of the traffic sent to Microsoft employees each day. This means that my figures for personal email volumes at Microsoft are distorted since they don't include junk mail deleted automatically.

It is not easy to quantify how busy all this email makes us. We are convinced that we are busy and that we receive too many messages. We assume that email is the main villain here, however many words we send and receive via our instant messaging accounts or other channels. Indeed, email overload (and the associated technologies that make that overload worse—BlackBerry devices, for example) is the main focus of our complaints. Many newspaper opinion pieces and entire books are now available on this topic. But just as I have problems calculating just how overloaded my colleagues are, so do these books seem to have similar problems. Christine Cavanagh, for example, begins her book *Managing Your Email: Thinking outside the Box* (2003, 2) with the curiously vague assertion that the "average daily volume of email is 50 per day," but she doesn't say whether this is for the entire North American continent, for the average professional worker, or for all users of the Internet. No one doubts that we email too much or that how-to books can be valuable (see, for example, David Shipley and Will Schwalbe's 2007 *Send: The How, Why, When and When Not of Email*).

In this chapter, part of my topic—even if it is a slippery one—is what we mean by *volume*, *amount*, and *overload*. Another part has to do with the idea that messaging is good for you.

Presumably this idea led society to make the technologies that induce us to make too much communication and hence to complain about overload. So part of my topic in this chapter has to do with a question that is prior to the one of volume—the idea that communication is good for you. Where did this idea come from? Who said it, when did they say it, and why? Just as it goes without saying that we have reached a point of communications overload, the idea that communication is good for you almost goes without saying. When my mother said, "Write home," I did not think that it was a good idea. But today, I almost doesn't query whether it is good to email to friends or to work colleagues. How did that change happen? Did my attitudes simply alter over time? They often do as you get older. If so, why? Or does the change have something to do with the differences between hand writing a letter and typing an email? I might look back and think that I didn't want to write home because it was too much effort, compared to how much easier emailing is today. But if the value (communicating) is worth only a minimal degree of effort, then how great is that value? Writing home with pen and paper was not worth the benefits that might accrue, but emailing is? Well, apparently yes.

Twenty years ago, the excitement was about methods—email in contrast to pen and ink (see, for example, Lee Sproull and Sara Kiesler, *Connections: New Ways of Working in the Networked Organization*, 1992). But today, the excitement is about new channels—blogs and Twitter being the latest revolutions in the age of digital communications. Cavanagh's book is but one of many on how to manage email, and there are at least as many books about what blogs can do. But the concern with email was simply about overload (about having too many emails), and the excitement about blogs is of a different order. The book titles say it all: *We've Got Blog: How Blogs Are Changing Our Culture* (Rodzilla 2002) or *Blog: Understanding the Information Reformation That Is Changing Your World* (Hewitt 2005). It's not merely that technology has changed, that we are fighting off a torrent of communications, or that someone persuaded us that communication is good for us (before we ended up creating a world where there is too much communication). We (according to the bloggers) seem to have changed, too, or we are being urged to do so. It's not just a question how much or who led us to this state. It is also a question of what we have become or ought to become. This too is part of the topic of this chapter—the *who* of communication as well as the *counting* of it, the *why* of it as well as the *what* of it.

A VIEW FROM THE PAST

The point of the opening remarks on my mother's admonishments was to note that understanding how we are now requires understanding how we thought of ourselves in the past. And yet doing so always entails a particular problem when it comes to our communicative practices: evidence is thin. Our understandings of ourselves in this regard are not necessarily based on facts and figures (volumes of letters sent or emails received). For one thing, measures of what we did in the past often have no equal measures today. There are no records of telephone or email usage before these devices were invented. Even after the telephone was popularized, most people didn't keep evidence of how much time they spent on the phone. Today, measures can be curiously misleading (as I say, the numbers of personal emails received are occluded by the vast volume of junk mail). Besides, in the past, it was not always clear that something needed measuring to document a change that was happening. By the time the change occurred, the time to measure it was past. This is certainly what Claude Fisher argues regarding the telephone (see *America Calling: A Social History of the Telephone to 1940*, 1992). Contemporary commentators worry about this, too. One concern is that the Internet might be creating changes that are not yet recognized and that when they finally are recognized, it will be too late to measure them happening. Only the consequences will be able to be measured (see, for example, Jonathan Gershuny's 2000 *Changing Times: Work and Leisure in Postindustrial Society* or his chapter "Conclusion: A Slow Start?" in Ben Anderson and colleagues' *Information and Communication Technologies in Society: E-living in a Digital Europe*, 2007).

There are some measures, and having figures with problems is surely a better starting point than having none at all. Nevertheless, figures rarely tell us all we need, even when they are fairly comprehensive. They always need to be understood alongside other evidence. In this case, the best approach might be to understand the figures (however good or bad they are) within in the larger landscape of the everyday formulas and commonsense frameworks (often apocryphal but consequential nonetheless) through which the world as we experience it is constructed and understood. These help give meaning and context to the figures. For sociologists, cultural theorists, and others, a narrative is made up of metaphors, facts, figures, and synonyms that somehow give labels and identities to our doings. These narratives can give each era its name or sense (one current narrative calls our era the network society). They are often used to describe and encapsulate the past, too. Without presenting arcane arguments about the relationship between these ideas and the measures of our actions (between cultural practices and bodily ability, which is a classic sociological chestnut), I note that these narratives are very much part of the way in which we account for ourselves and the society we live in. They are bound up with the society we are part of and also with the ways that we frame ourselves and account for what we do. They help us count what it means to be human. To understand what it means to be human at any time therefore entails tracing the ways various narratives describe, account, and explain that humanness. In the example above of my mother's admonishment to write letters from college, her present was bound to perceptions (wrong or right) of how some past had been. What was that past, and can we learn anything about ourselves by looking at it?

From my mother's remarks, one imagines that letter writing was common in her past. More than this, her remarks imply that letter writing was a virtue of some sort. What she said also implied that the virtue was being lost even as I walked out the door. (Her failure to mention telephones reflects her frugality and the formerly high cost of long-distance telephone calls.) She was not alone or even unusual in conjecturing about these sorts of things. Her understandings, like those of most of us, came from many prosaic and commonplace sources—facts and figures, ideas about cultural values, and notions (perhaps snobbish ones) of what it was to be sophisticated, educated, and so on.

My own views on letters and the art of letter writing were constructed at school, where I was presented with historical examples of great letter writing as part of my history and English literature classes. In our history lessons, samples of letters were often presented to conjure up a sense of place that might otherwise been difficult for the teacher to convey. In English literature classes, letters were presented as examples of style, form, and narrative. These letters were varied but of two general types—letters that seemed to be forms of art and letters that seemed to be about the machinery of life. Both, in different ways, led me (and I think others) to construct a vision about the people who wrote letters and the circumstances in which they lived. Through the use of these letters, I was taught a sense that I existed in a culture that emphasized writing and sending letters as a special or certain kind of cultural act. The first type of letter—a form of art—is exemplified by those compendia of letters that detailed the lives of eighteenth-century English aristocrats. Individuals like Horace Walpole wrote letters as celebrations of their grand tours as they lavishly traveled around Europe. On the grand tour, the young and well-heeled visited the ruined delights of the ancient world in Rome, Athens, and elsewhere. These letters were not simple travelogues itemizing "I did this, and then I did that" but were written with great craft to display the insights, thoughtfulness, education, and knowledge of their authors. Moreover, they were not written merely to friends and intimates, although they were addressed to them. They were written in full knowledge that copies would be collated and sold in bound editions on the sender's return. As a case in point, Walpole wrote his letters not just to tell his story to his friends but also to make money by selling copies of those letters afterward. His letters were a business venture designed to make money out of self-celebration. Writing letters was a way of climbing the social ladder in eighteenth-century London.

In the nineteenth century, letter writing had a different hue, with the production of them as "grub street publications," as

Samuel Johnson so dismissively put it, being less of a concern (i.e., to make money). But people still wrote delightful letters that are works of art. William Gladstone, a British prime minister during the second half of nineteenth century, sometimes wrote twelve letters a day, some over twenty pages long. Some of these were collated and published, mostly after his death, with a view to honoring the man, not make him money. His letter writing was not broadcasting in the manner of Walpole, but the letters certainly were a manifestation of an articulate person and his artfulness, an art that was constrained by the properties of the written word on a piece of paper either folded on itself to hide its contents or enshrouded in a bag (an envelope).[1] Not everyone will be happy with the word *art* here. I remember thinking that Gladstone was a man who liked the sound of his own voice more than anyone else's, and some of the letters seemed too long. Some aspects of the letters were curious. One learned about the weekend house parties and the ways the sexes were separated. Gladstone would write about the things discussed with his male friends once they had "withdrawn from the women." Cigars seemed commonplace, too. They evoked an atmosphere of ponderous conversations, heavy with the scent of furniture polish and smoke, of masculine vanity and feminine invisibility.

What I was being instructed in and was fumbling toward was an understanding of *cultural practice*. This particular practice has been written about a great deal from two points of view. On the one hand, eighteenth- and nineteenth-century instruction manuals outline what letters as art are meant to be. They present narrative lengths, sample introductions, stylistic patterns indicating the voice of the writer (as dutiful husband, child, or courter), and much more. On the other hand, many studies by historians describe the emergence of the cultural practices in question. Although many of the instruction manuals were British (such as Samuel Richardson's *Familiar Letters on Important Occasions* of 1741; Richardson is also the author of *Clarissa*), the most lively historical commentaries seem to be American. Many of these historians view letters as a key part of America cultural history.

There were various similarities between the letters of, say, Walpole, and American letters of the same era. One similarity relates to their form. Their style was formal, rhetorical, essay-like, and even moralistic, which has led to a view that these kinds of letters are epistles of sorts, as in Paul's letters in the New Testament. Although there was little religious intent in the letters of Walpole or Gladstone (even though the latter was an earnest Christian), their style made them seem like epistles. They were almost oratorical, dealing with particular topics presented in a narrative form, and they had length: an epistle could not be only two lines long.

As this style evolved, letters began to encompass a broader range of genres. Predominant among these was a form that began to emphasize the intimate and that created a new engagement between sender and receiver. This was especially so in the United States beginning in the mid-1850s. Most American letter writers of this time were not indulging in aristocratic travel, as Walpole did a century earlier, but their economic, social, and geographic locations (including the western Gold Rush and the Civil War in the south) created a setting in which a "postal culture" began to emerge. As David Henkin suggests in *The Postal Age: The Emergence of Modern Communications in Nineteenth-Century America* (2006), at the start of the nineteenth century, letter writing was key to the spread and development of business, and by midcentury, letters and letter writing became a key cultural mechanism of American identity:

> As a newly accessible and increasingly indispensible communications network took root in the 1840s and '50s, American correspondents sought to articulate new models for postal relationships.
>
> A great deal of cultural work went into the production of the codes and ideals of intimacy that shaped epistolary communication. What emerged most generally . . . was a set of practices, discourses and beliefs—a postal culture—that redefined the very status of mail. (93)

What these writers invented was the personal letter. It was personal not in the literal sense that it was addressed to someone but in its tone, manner, and topic and in the sensibility that the sender and the recipient had toward one another. Although letters might be written between persons of very different social status (a father to a son, for example), they were constructed to convey a sense of a special moment together between the sender and recipient, like a whispered but lingering intimacy in a private room. This intimacy gave the letters their charm and made letters precious.

There were various consequences of this evolution. For one, the privacy of letters became of utmost importance. Despite the dozens of hands that might touch an envelope, its seal was viewed as sacred. It still is. Above all, what was inside the envelope really encapsulated this style. The style was not learned overnight. It took a great deal of effort and instruction: courses were offered, books written, and style codes shared round. Success at producing personal letters—at having the artfulness to do so—was highly regarded because it required considerable finesse and facility with styles and form. Moreover, these styles could not be adopted just by a close reading of the manuals. In the letters that they wrote to one another, Americans taught each other "how to see and recognize letters as gestures of self-expression" (Henkin, 2006, 117). These gestures became measures of cultural finesse.

This was not an isolated moment in history that has no consequences. What happened then affects us now. Henkin explains that the sense of this intimacy—the wonder it bestows and the delight it affords—has become an essential component of our contemporary existence:

> Despite all the changes that separate us from the postal culture of the mid-nineteenth century, our pervasive expectation of complete contact, of boundless accessibility, actually links us back to the cultural moment when ordinary Americans first experienced the mail in similar terms. The world we now inhabit belongs to the extended history of that moment. (175)

This is a bold claim. He suggests that as we sit down at our PCs and key some new notes to our IM buddies or post blogs on our social networking Web sites, we are doing in modern garb what others were doing at the time *Moby-Dick* was written—namely, bringing ourselves together in a way that defies space, time, and physicality.

TIES BETWEEN WHOM?

Whether or not this last claim is true, what Henkin describes as bestowed on those who create letters and on those who receive them is a kind of virtue. This virtue has, on the one hand, something to do with the relationship between the letter and the writer, with the transforming experience that an individual goes through by dint of writing a letter, and on the other, something to do with how the receipt of a letter alters the character of the relationship between the sender and the recipient; it can also alter the sense of the self of the recipient. They may feel altered by having a letter written to them—honored, for example. Letters somehow let the participants transcend time and space—and not merely in a geographic or clock time sense. Letters bring people together in a new way, a powerful, expressive and evocative way that is not possible without them. Letters also alter the participants involved. Letters are not an analogue of face to face communication; they create a new experience of human bonding.

What is perplexing here is that making bonds appears to be only part of the virtue in question. After all, one of my problems as a teenager was that I did not want to create ties with my mother. Letters had no appeal if that was their achievement. She wanted to preserve that tie and even deepen it as I got older. I wanted to break it. Like all teenagers, I wanted other ties, even if it came at the cost of the one with her.

Yet even if I was willing to write letters—to make the product that could create ties—I had another problem. As I walked out the door and headed off to university, my concern was not that I could not write a letter but that I might send letters to those who did not want one. I was worried that my efforts to make ties would end up not in written conversations that bind but in soliloquies that humiliated me. Who would I write to? Would they reply? What would induce them to correspond with me—my spotty, undesirable, awkward self?

So a further facet of the virtue that is associated with letters is the fact that someone has others who desire their letters. The paradox here is that letters can create ties only when those ties somehow already exist. And beyond this, the ties that they create—on the basis of a modest but nevertheless marvelous transcendence of time and space—is of a different nature than the ties that allowed the correspondence in the first place.

Historians like Henkin don't much mention the relationship that might need to exist between people for letters to do the work that he says they do. Those relationships are taken for granted in their studies. The problem of who participates in such a relationship (whatever the media or channel) perplexes us now much more than ever before. If letter writing was developed as an experience between intimates, what is the experience of blogging about, for example? Is there a connection between the evolution of letter writing and the current passion for one-to-many letter writing (which, in a sense, is what blogging is)?

I am highlighting these facets of what personal letters achieved and required (which I label as a virtue) since I want to take seriously Henkin's claim that, in our current circumstances, we are extending the sensibility in question. But his analysis doesn't tell us all we need to know. We can discern some (but not all) residues of the desire for personal letter writing in our practices. There are other aspects to our motivations that lead us, now, some 150 years later, to be willing to communicate, exchange, and transform ourselves even with strangers. Mid-nineteenth-century Americans learned to savor letters, but most (if not all) of these letters were between souls known to each other. Today, many of us seem to delight in creating ties with those we have never met, and we do this on a large scale. Personal letters once created a bond of written intimacy between persons who in most instances had close ties (or whose relationship would become closer by the use of the letter), but today, we use an evolved form of the personal and intimate style to create moments with those who are foreigners to us. Blogs are constructed as if they are offering some of the private remarks and experiences that used to constitute the stuff of personal letters, and yet blogs are written for the digital crowd of thousands of strangers. Between these types of letters—the narrowcast and the broadcast—many other modes of expression are facilitated by social networking sites. These too create different manners between people. Facebook offers an experience that entails letter writing to people one knows (and hence is unlike blogging), but its status alerts and message posting have aspects of a public performance (unlike the letters that Henkin considers).

BLOGS AND EXPRESSION

I am marking out a connection between the writer and the reader. Various commentators have suggested that the Internet is transforming the form that communicative relationships have. While social networking sites are doing this in one way, other forms of mediated communication are doing it in other ways. These changes are affecting relations between people, shifting the landscape of human connection.

Letters bestow a virtue and deepen the ties between humans, and blogs also provide a virtue and foster human relationships. Let's approach these two virtues by looking at a particular social relationship—the one between people and organizations and the people those organizations serve. Apparently this relationship is changing and becoming more humanlike. Numerous authors claim that the formerly characterless nature of organizational communication is being replaced with something new: a form of communication that is celebrating "the human" in organizational life. This new style of organizational communication allows people to express more than drily neutral information about corporate actions. Instead, employees are expressing their delight or indignation about those actions and conveying their moral judgments. According to many commentators, the blog revolution is bringing back passion.

Yet when they use such hyperbole, they are simply referring to what might be the most anodyne of all formats—an inversely chronological Web log. This is a Web site in which the most recent entry (usually textual) is added topmost. What art in this, one might ask? What communicative wonders does this achieve? Blogs could simply be the text that people choose to post onto their own personal Web sites each day—texts that are essentially digital versions of the traditional diary. But they are not this. Nor are they like letters written between two persons who are close to one another and seek a sense of intimacy in the production of the content. They are something different from this, too. So what is this mode of expression? The advocates of blogspeak have opinions on this matter (just as those entranced by the power of the handwritten letter had strong opinions on what that mode could do when it was revolutionary—say, in the 1850s). Let's take Robert Scoble and Shel Israel's *Naked Conversations: How Blogs Are Changing the Way Businesses Talk with Customers* (2006). In their view, blogs are supposed to be written from the heart—to be produced passionately rather than dispassionately, to be off the cuff rather than planned. Above all, blogs are meant to be better than marketing corpspeak:

> Bloggers . . . are generally suspicious of the smooth and refined language of official spokespeople. They use terms like "suits" to imply a suspicion that there is no human inside. Spokespeople use a strange language called "corpspeak," an oxymoronic hybrid of cautious legalese seasoned with marketing hyperbole. (4)

Proponents of the blogosphere, like Scoble and Israel, claim that blogs offer a corrective to the bland and not always frank words of corporate communication. People inside organizations can blog on behalf of organizations and in so doing change the form of the relationship between the corporation and those outside it. They believe that bloggers on the inside communicate in a fashion that makes the corporation have a human face and be seen to have people who are prepared to be accountable for the actions of the corporation. So powerful is this benefit, blog advocates claim, that all organizations will need to allow and even encourage bloggers to blog from within their walls to give that humanizing, accountable face to consumers on the outside. My own corporation certainly views communication this way and sees the emergence of blogs from within its corridors as giving a human face to a corporation that had hitherto been treated as an anonymous and somewhat threatening corporate monster. Scoble and Israel were, in fact, Microsoft employees when they wrote their book.

THE TOPICS OF BLOGS

Blogs are not to be thought of as honest, passionate soliloquies, though. Blogs have properties beyond the identity of the blogger being visible. For one thing, blogging has become a mass phenomenon. Yet blogs are not quite pronouncements that are broadcast or consumed en masse, like TV news. They tend not to be diligently prepared reflections on particular topics. Although blogs can be written about any topic, they tend to be characterized by their spontaneity. They are (to put it in blog speak) splurges on matters that concern bloggers as a whole, as a gang—although some bloggers tend to lead the blogosphere by opening up new topics and concerns.

This gets to the heart of why blogs are not diaries. The topic—not the passing of time (as with diaries)—drives their production. Blogs are not read randomly but through a gradual process of selection and monitoring. Blogs are read as a web of interrelated objects that are connected through systems of *really simple syndication* (RSS). These produce *feeds* (mentioned, it will be recalled, on the first page of this book) (pieces of information) that are pushed to those who are in the syndicate. Unlike business syndicates, the number of subscribers is not limited or small, and most Web browsers allow users to add an RSS feed whenever they wish to do so. Syndicating can be enormous, even though few blogs generate feeds that lead thousands to read them every time that they are changed. The blogs are experienced virally. Blogs also push themselves onto the computer, encouraging their readers to keep up or, if they have been neglectful, to catch up.

In the *blogosphere*, people are reading and writing blogs every day because they are researching every day. Even if one does not produce a blog entry, others do, and the blogosphere becomes a universe of its own, with daily topics, threads, and lingering concerns. Not everyone participates equally and continually; some people indulge only occasionally. But there is a sense that what people experience in the blogosphere is analogous to the touching of a heartbeat—the human reflection on the topics of the moment. No one person owns this heartbeat or produces it; it's a collaborative affair. The value of searching through the webs of RSS feeds and understanding the threads is to be able to add comments to those threads. Only in this way does the pulse remains vital. To know what the action is and what is being said about the action is to be in the blogosphere. Keeping up is what makes the blogger.

For some people, this experience is addictive, and being in touch is a never-ending search. In a Kierkegaardian irony, as soon as they know what the latest thing is, the knowing of it makes it no longer as interesting as it once was. Newness is where the action is. But how much time do bloggers devote to these activities? How do they manage to be in touch without compromising everything else that they do? For most people, practical constraints affect this, and besides sometimes there simply isn't much news to be had. Even the most avid bloggers find that on some days there is nothing much happening.

This leads us back to the question of numbers. Perhaps here numbers and countings of communication events might help. Perhaps arithmetic is an index of the blogosphere, even though the bloggers might think this to be inimical to their rationale. At this point, for example, I could list the number of RSS feeds being set up each day, the numbers of Web sites that these attach and link, and the broadening universe of bloggers worldwide. Even as I write, however, the numbers expand: millions now blog (though not all blogs are read). If we are to believe Scoble and Israel, the numbers will continue to increase. There are also software applications called *aggregators* that aggregate feeds to distill their contents for the overworked blogger. This technology is designed to solve a problem caused by a prior technology (RSS). But there is a moral here. Sometimes we design a new communication tool only to find that it consumes time and hence needs controlling, and so we invent *another* technology to do that controlling. Where might this end? We have reached back to the opening remarks of this book—the paradox that lies at the heart of our current world. What we invent creates problems, so we create solutions, but these aren't solutions so much as temporary fixes for a problem of our own creation—too many means of communicating too often.

MEASURING PEOPLE, MEASURING TIME

Blogging is only one species of communication among many. It may be thought of as one-to-many communication, (a kind of broadcasting), whereas other kinds of communications are one-to-one (emailing or texting between friends for example). Between the pushings and pullings, these narrowcastings and broadcastings, these circulations and sharings of information, the occasional bespoke missive, are various other kinds of *communication act*, including instant messaging, tweeting, and even old-fashioned letter writing. If blogging is about keeping up to date with the hubbub of the moment, then an email or a text between friends is achieving something different, a turn at a conversation, a move that creates a sense of intimacy between those involved. Similarly, although we have not addressed IM in any detail, an instant message assures an intimacy between the participants but might be more ephemeral, with each message requiring another soon thereafter to keep the intimacy vital. Each type of communication act achieves something different.

One might presume therefore that the value of each communication act ought to be judged in different ways too. But judgments about communication overload most often talk about time—about how much is given to reading emails, texts, blogs, or instant messaging, about how much time is left to do other things, for example. It is certainly true to say that time is a kind of constant here, but just how useful it is as a measure of the value of an act of communication is another question. Is the time given to keeping up with the blogosphere to be measured as equal to the time given to sending a letter? Isn't one of the differences between a blog and a letter that one is created off the cuff and in the passion of the moment whereas the other is planned and thought through before it is posted? Whatever the amount of time one gives to each of these tasks and irrespective of their value, it is certainly true that there is only a finite amount of time available.

Let me begin to unpack this problem of time by returning to the letters that I read at school. In addition to the artful letters I also had to read letters about the machinery of living. These letters came from housekeepers, servants, farmers, and laborers and listed jobs to do and budgets to spend. These letters gave us students insights into the lost details of the past—as a way to get a sense of the endless chores, for example (cleaning fireplaces, filling up coal bunkers, doing laundry). Some of these letters were imbued with moments of artfulness and with asides about feelings and passions (aching over a separation was common). As I read these letters and reflected on how the world might have been, I compared that world with my own life circumstances. Despite my own insouciance to practical affairs and despite evidence to the contrary, I read these letters as proving that I was busier than people in the past must have been: "They had time to write letters. How different my life is!" I was sure that I was too busy to write letters (although busy doing what, I was not quite sure). The existence of letters was proof to me that the past was a quieter, more peaceful place that was more conducive to letter writing than my own time was. Despite the contrary evidence in some of the letters themselves, I saw this legacy of letters from the past as indicators that in the present world that I was about to enter everyone was much busier and more earnest in their desire to work and to play than those past letter writers had been: "Today one cannot write letters because one has too many other things to do!"

Was this view correct? My conception of the past was made up of assumptions and prejudices, naive opinions and curious facts. My visions were hardly good history. Whether my understanding was good or bad, however, what I thought was pretty much what many people thought at the time. This was the world as most people perceived it in the 1970s and 1980s. Was that picture really true, though? Perhaps everyone was busy in the '70s and '80s, but were people who lived in the nineteenth century and earlier really *time rich*?

The Royal Mail (and various other institutions) provides data that offer answers to this question. Simple contrasts between mail usage then and now are difficult because there were lots of changes in the patterns of mail usage in the nineteenth century. One important innovation was the Royal Mail's introduction of the Penny Black stamp in 1840. The Penny Black was the world's first generally issued prepaid adhesive postage stamp, and it allowed people to send a letter anywhere in the United Kingdom for a standard fee. Before the Penny Black, letters were paid for by the recipient on delivery, and the cost depended on distance traveled, difficulty of locating the recipient, and size and weight of the letter (although there were no standards and measures in this regard). The recipient basically paid a negotiated fee for someone else's letter. After the Penny

Black, all the recipient had to do was be thankful. Letter writing became a much larger phenomenon not because people had any greater desire to write and express or because they had more time to write and read but simply because postage was more practical and much cheaper.

After the introduction of the Penny Black and throughout the remainder of the century (the period that Henkin writes about), however, one does not find that heaps of letters started dropping on door mats. A lot of letters were sent, but per capita this did not turn out to be many. Our perceptions of letter writing in the past—its volume and its importance and centrality to people's lives—are not very close to what actually happened.

Table 2.1 summarizes the important features of mail volume in the United Kingdom from 1840 to 1910. The remarkable growth we see in this is also a reflection of the equally remarkable growth in population during that time. British population grew from 18 million to 41 million, while letters increase from 151 million to over 3.6 billion per year. So people went from receiving an average of eight letters (and postcards, which were invented in this period) per year to receiving about twenty-nine letters per year. The numbers can be increased a little if we look at only people who were "letter literate"—the urban, affluent, and educated. For these people, between 1840 and 1910, the volumes of letters and postcards received per capita increased to about one hundred and thirty per year. Thus, in the heyday of letter writing, the letter literate received only about two per week (and this includes all sorts of letters, not just epistles).

Some households would expect a letter on most days and perhaps would wait in anticipation for the postman. But this was a delight that only the elite letter literate enjoyed and was

TABLE 2.1
Estimates of mail per head in Great Britain, 1840 to 1910

	1840	1850	1860	1870	1880	1890	1900	1910
Population:								
Total population (mil)	18.3	20.6	23.0	25.8	29.4	32.8	36.7	40.5
Adult population (mil)	11.9	13.4	14.9	16.5	18.8	21.4	24.8	28.0
Adults (%)	65	65	65	64	64	65	68	69
Literacy:								
Adults who could read (mil)	6.9	8.6	10.7	13.2	16.1	19.9	24.0	27.2
Men (%)	66	69	74	80	86	93	97	97
Women (%)	50	60	70	80	86	93	97	97
Adults (%)	58	64	72	80	86	93	97	97
Adults who could write (mil)	4.6	6.0	8.0	10.6	13.7	17.9	22.8	25.8
Adults (%)	67	70	75	80	85	90	95	95
Items mailed per year:								
Letters (mil)	151	311	516	783	1,052	1,553	2,106	2,774
Postcards (mil)					109	207	384	831
All mail (mil)	151	311	516	783	1,161	1,760	2,490	3,605
Items mailed per literate adult per year:								
Letters	33	51	64	74	77	87	92	107
Postcards					8	12	17	32
All mail	33	51	64	74	85	98	109	139

Source: Data from Hamill (2008).

probably further limited by scope of delivery (urban centers guaranteed several deliveries each day). But letters and letter writing did not consume a great deal of time, even for them. The golden age of letter writing did not, it turns out, involve much letter writing for most adults (even if on a societal level millions of letters were produced and had to be delivered). Although the culture of the time encouraged people to delight in and honor this form of expression, people would not have complained about communications overload. They were right not to. They weren't overloaded.

TIME, LETTERS, COMMUNICATION

But frequency of letter writing might not be the only dimension that should be considered. We might approach letter writing in terms of how much time was available for their production—as remarked, when I was a teenager, I believed that there was more time available in the past. In my view, those in the past who wrote letters simply choose to do so from a list of alternative ways of whiling away time. But did they? Or was the situation more complicated? Did they have to work at making the time to write?

One problem with answering these questions is that measurements for daily tasks in the past are far more gross than the measurements we have now (today we have studies that examine how daily lives are scheduled into periods for work, rest, play, TV watching, Internet use, and so forth). It has been suggested, for example, that between 1870 and 1979 (a bit more than a century) the amount of time that males spent working halved (Hamill 2008).The same author also argues that we spend about one third of the amount of time on work that people in the Middle Ages did (see Hamill 2010). Such figures are broad and ignore gender, domestic work, and the ways that some activities are interleaved. Indeed, this issue perplexes current time measurers, who have sought various means to measure polychronicity. Success here should not distract us from fairly universal sources of time consumption, such as sleep. These and other biological needs account for about half of all time usage.

These days, we have a great deal of research that counts up our daily minutes, even though our multitasking presents problems for these studies. In addition to the large trends (particularly the move toward less work and more leisure) that they reveal, these studies present some curious facts (such as the numbers of letters written in the past were not as many as one imagines).

For example, one learns that people spend on average about 8 percent of each day communicating (Hamill 2008). That percentage seems small—given that each Microsoft employee receives about 120 emails a day each (even subtracting the filtered-out messages, the number of messages that they have to deal with is likely to be fairly large). Many Microsoft staff members also use instant messaging and receive and make phone calls, as well, so fitting all this communicating into a figure of 8 percent suggests a remarkable compression of communicative effort into very small amounts of time. If these figures hold true, then one can imagine my colleagues having only moments to deal with many emails, IMs, phone calls, and texts. My colleagues must have remarkable facility with time management.

The problem here is that some figures suggest one thing and other figures suggest the contrary. Volumes of messaging undoubtedly go up enormously, but measures of the amount of time consumed by dealing with messaging don't seem to reflect that. In 2005, nearly half of the UK population used the Internet daily, for example. According to some assessments, this means that the modal user received only five personal emails a day. Meanwhile, people spent about eleven minutes on the phone (fixed and mobile) and sent about a three texts a day. In other words, the figures for technologically mediated communications are low. Is it possible that just as we felt that many personal letters were written in the past (but didn't know that this was entirely true), we claim the same regarding the Internet? Do we think that everyone communicates a great deal, but some may and most don't? Are we being mislead by the vast numbers that aggregates of messaging volume produce?

It's not quite as simple as this, although communication is up for everyone—even if we can't be quite sure how far up is. At a population level, trends that seem to be important don't apply themselves evenly. Better studies are needed of those who message and those who do not and of those who have adapted to the communications age and those who have not. We need measures of, say, how much messaging teenagers and twenty-somethings do rather than numbers for all adults.

EXPRESSIVE COHORTS

Gershuny, in the somewhat dated book *Changing Times* (2000), claims that communications are unevenly distributed. He suggests that the more education an individual has, the more time they spend at work online and hence communicating. They communicate more because of their status, he says. If this is true, then my colleagues and I might simply need to acknowledge that we are busy because we spent too long at university. Our doctorates have led us to work long hours and consume our lives communicating to a web of fellow sufferers—the overeducated and hence *overworked* classes. Gershuny's thesis is delightful and has some evidence supporting it, but it also has generated some rancor. He is essentially saying that the relationship between social status and work has inverted in the past fifty years or so. Before World War II, blue-collar workers (those without professional qualifications) worked more than workers with high school or university degrees. More recent studies show that this difference in educational or cultural type is manifesting itself in a divide between those who are displacing their television watching with Internet use and those who continue to absorb themselves in broadcast media. According to William Dutton and Ellen J. Helsper (2007, 24), by 2007 nearly half of the UK adult population used the Internet, thus shifting their habits away from television consumption (although those who turn to the Internet seem to have watched less television anyway). For those who were making this shift, over eleven hours per week were given to the Internet and sixteen to television, while those still consuming TV spent twenty-four hours watching shows like *Big Brother* and *Friends*. For those who were turning to the Internet, the primary activity was email, and e-commerce was second.

The figures for Internet and TV activity suggest that a remarkably large amount of time is given to these activities, considering total amount of available time (when sleep, eating, and other prosaic activities are considered). But other studies show that children and teenagers are spending startling amounts of time engaging with the Internet. According to Amanda Lenhart, Mary Madden, Alexandra Rankin, and Aaron Smith in their *Teens and Social Media* (2007), 93 percent of U.S.

twelve to seventeen-year-olds use the Internet, and of these young people, 64 percent do so with content creation in mind (such as emailing, using instant messaging, and posting content on Facebook or MySpace). Moreover, when they post images or videos on Flickr or YouTube, their purpose is to provoke communication—emails, IMs, and Facebook comments. Finally, 55 percent of this age group create profiles on social networking sites—not simply to have a digital profile but to use this to anchor and foster digital expression.

Although these individuals give a large amount of time (indeed, an ever increasing amount of time) to digital expression, the scale or range of their social contacts is not also increasing. As danah boyd has noted in "Taken out of Context: American Teen Sociality in Networked Publics" (2008), teenagers use social networking sites and even YouTube not to express to the mass but more often to express to those they already know well—often a handful of friends. Although the digital world might be made more intense by digital networks, by this measure it is not being made bigger.

One might add that although teenagers might only communicate with half a dozen friends, they are always wary of the possibility that their postings on Facebook and other forms of self-expression might be released and broadcast to the digital crowd. Even though they use the Web to whisper and intrigue with one another, they need to handle the new medium carefully to avoid having themselves become blog news for hundreds.

ORDINARY DOINGS ON ORDINARY DAYS

One thing that time-measurement studies reveal is how little time we have. When all tasks are added up, we spend about

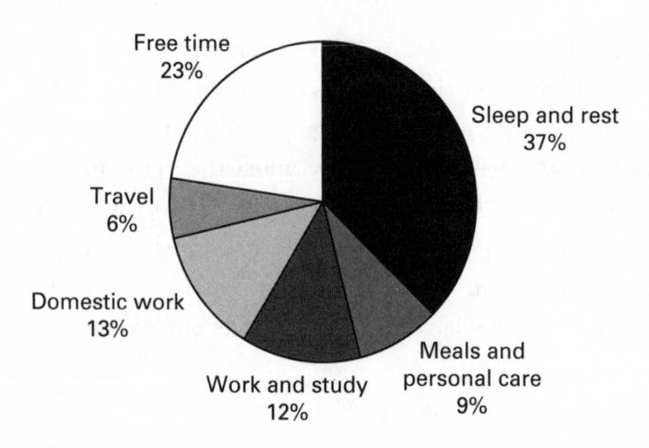

FIGURE 2.1
Division of the day in the United Kingdom, 2005, calculated by averaging time consumptions for all individuals during a one-year period
Source: Office for National Statistics: The Time Use Survey 2005, HMSO, London.

37 percent of our time sleeping and resting, 9 percent eating and washing, and 6 percent going to and from work and undertaking other errands (see figure 2.1). About half our lives are consumed with getting ourselves ready and fit and in the right place at the right time, and from what is left, we do paid work and domestic chores. Only then can we start thinking about letters, email, texts, and instant messages and keeping up with Facebook.

Now, it has to be admitted that these figures might be misleading because they are calculated by averaging activities for all working and nonworking adults over a year. Hence these suffer from the same problem that the figures for Internet traffic do: they are too aggregated to be helpful unless one is very guarded in their use. Just as cohorts that communicate are likely to be different (teenagers and older people, for example), their time habits are likely to be different. There don't seem to be many statistics on those differences. Teenagers doubtless do communicate more than older people and are likely to spend less time on domestic chores than older people. But the figures can't be marshaled to prove this.

Though we can use our family experiences to assert the fact that teenagers spend a surprisingly large amount of time communicating we can use the same family experiences to admit that it is not they who are complaining about communications overload. The older cohorts (where the evidence is far less persuasive) are. Quantitative measures are not driving the issue. Countings are not leading to complaints or at least not in a simple way. Perception must be involved.

One of the possible sources of this discrepancy between doings and complaints might have to do with differences in the kinds of tasks that different cohorts undertake. One of the familiar complaints one hears within families is that teenagers don't do their share of the housework. Another is that parents don't believe that teenagers can do their homework while using social networking sites. Teenagers often respond to these doubts by saying that they are using the social networking sites to talk with the classmates about homework. It is not distracting but helping them. Like parents, time measurers have difficulty with polychronicity—the fact that people do more than one thing at a time.

For instance, the proportion of time given to watching television might seem remarkably large. According to some estimates, it is about half of the leisure time available to the average TV user in the United Kingdom, for example. But this does not allow for the fact that television is not a monopolizing activity. One can eat and watch TV, one can talk and watch TV, and one might even have TV on as wallpaper (in the background) when one is doing nothing. The data might say that the TV is on half of the free time that people have, but that doesn't mean that they are watching it. They might not be suffering from TV overload, even if the figures suggest that their lives must be saturated with soap operas. Despite the best efforts of the time measurers, some facts about human life seem to be beyond their reach. All too often, people don't concentrate on or provide undivided attention to one thing at a time, a fact that often eludes researchers.

It is not that the researchers are unaware of polychronicity. The methods that they use (diaries, lists, categorization of tasks) engender views of the world that cut it up in simple ways. To make things worse, the subjects of their inquiries are not entirely helpful. Keeping a minute-by-minute log of what we do is not easy. How often might one speak to one's partner while watching TV, for example? How many conversations might one have when the TV is on that one does not bother to report on the log? One might complain about the data, but one can understand the all too human reasons for its poverty.

DO WE HAVE ANSWERS?

There might be a difference between our complaints and certain quantitative measures, and these two should not necessarily be linked in a straightforward way I am suggesting. I have sketched out, too, a number of themes and perplexing facts—setting the scene for the final step in our inquiries.

First, for many reasons, it can be difficult to ascertain how busy we are. It is certainly hard to get a sense of how much communication we do if we confine ourselves to figures. We have gross figures about life at work and sociality at home. We have seen that those who seem to communicate the most and who seem to be most afflicted by communications don't seem to complain about it. Teenagers don't bewail the fact that they have too many messages. The real issue is that for most adults, there is a *perception* that we are now suffering from an age of communications excess. This is constructed, in part, by seeing the past in a particular way. Whatever our current circumstances, we tend to believe that the past was different from our present. We complain that we are busy and overloaded (and one source of that overload is communication), and we therefore portray the past as less busy and not like our today.

But we also use that past—in this case a real past and not a fictionalized one—to help construct the sense, purpose, or delight in mediated communication. This delight is one of the main sources of our busyness today. What motivates us now might have its roots in what motivated our forebears. The work of historians of postal systems reveals that communication is a special kind of act. As the written letter evolved into a particular form, a special relationship between sender and receiver began to emerge. The written word was orchestrated to allow the sender to offer advice, thoughts, and reflections and to lead the recipient to honor, esteem, and feel intimate with the sender. Letter writing was artful in its form of expression and also in its creation of a relationship between the involved parties that was intimate and that transcended time and place. The letter could allow a communion of souls—whispered intimacies, charm, private moments between participants, a sense of boundless accessibility. One consequence of letters was a deepening of the relationship between sender and receiver. Given this, it seems hard to explain other forms of communicative practice where the creator and the reader (synonyms for sender and recipient) are unknown to each other (such as the mass audiences for blogging). Blogging and social networking sites do not offer the same experiences as person-to-person communication like email or text does, but using them certainly consumes time.

The long and short of it is that one cannot really prove that a specific number of messages produces overload. Those who seem to message the most aren't the ones who complain about overload. Those who do complain don't seem to be sending large numbers of emails, receiving vast numbers of text messages, or entering posts on their Facebooks. But those who complain might nevertheless have reasons to do so, and this might have to do with the values that they apply to their lives. Teenagers might do nothing else but express on Facebook. As teenagers, they can get away with it (see, for instance, Angela McRobbie's book *Feminism and Youth Culture*, 2000, particularly the discussion of bedroom culture). But others have to work.

Who are these others? The author of *Managing Your Email*, Christine Cavanagh (2003) and Shipley and Schwalbe, the authors of *Send: The How, Why, When and When Not of Email* (2007) all seem to imply just about any busy professional person. For want of a label one might say knowledge workers. At this particular cultural moment, it is knowledge workers who are looking at their lives (and at the world around them) with a certain perspective that emphasizes some things at the expense of others. They look at emails, text messages, and postings on their social networking accounts and wonder whether they ought to respond to them immediately. And as they reflect on this they also hear on the radio, see on the TV, and read in the papers that society as a whole is suffering from overload and that communications are thought to be a prime source of this. No wonder then that they think their complaints apply to all, and that there is some objective proof that their email volumes, their texting, their social networking activities are indeed too great and that society is as a whole, overloaded.

Sociologists might say that this effect has a narrative—a set of ideas that lead people to think this way. They haven't labeled this particular narrative yet (the term *network society* was first used ten or so years ago, and some are now using terms like the *mobile society*) (see John Urry's book *Mobilities,* of 2007) as a case in point, but not anything to do with communication. Castells, who made most of the term network society in his book of the same name in the 1990s (*The Rise of the Network Society*, 1996), has hardly come up with a ringing replacement phrase with his latest book, *Communication Power* (2009). But regardless of labels or whether or not we think of ourselves as knowledge workers, nearly all readers of this book, just as I did when I commenced researching for it, will concur with this narrative. When we look at the emails, texts, and letters that are in front of us, we treat them as conspicuous proofs of what irks us. They become both a celebration (of being in touch and having friends, colleagues, and partners) as well as a burden that we resent. Our messages *and* our attitudes toward them are a measure of our age.

This explanation returns us to the paradox that is at the heart of this book. When we complain about the thing that gives us much of our contemporary self-esteem, it is not because we are false in our perceptions (that the numbers don't fit the facts) but rather because this is how we have come to see the world in which we live and hence make our judgments about it. The point is not simply about numbers or countings: it is ways of seeing and understanding.[2] Consider the Ottoman clock, which had twelve segments in each half of a day—one half of the day being defined by daylight and the other by night. The Ottomans used the actual sunrise and sunset as the demarcation between the two halves, irrespective of the changes that the season made to the rising and sinking of the sun. In the summer, the hours of daylight were longer than the hours of night, and in the winter, the hours of night were longer than the hours of daylight. Visitors to the Ottoman court in Constantinople found this confusing, but the Ottomans themselves thought it entirely rational: it reflected the value that daylight and night time were given. There was less time in the winter and more time in the summer—by which they meant more time to do the things that mattered (such as building, traveling, and making war). Similarly, because we don't give as much time to communicating as we do to, say, sleeping, we should not think that we treat these two activities equally. Indeed, it is evident that we don't treat them as the same. We might spend ten or fifteen minutes dealing with three or four emails each day, but this time is not equivalent to ten or fifteen minutes of sleep, washing, or doing domestic chores. They are longer minutes than the ones we give to sleep, as the Ottomans might say. And when we do communicate, we experience that delight, wonder, and magic that Henkin pointed toward—that sense of togetherness with another, a transcendence of time and place. The minutes that we give to communication, even though they might be few, are the minutes we honor above many other things. They are the ones that count. It is no wonder, therefore, that we complain when we find that we are using them up and when the thing that causes us to consume that precious time is the very act that makes them precious in the first place—the communicating, the keeping in touch.

A NARRATIVE FOR THE COMMUNICATING HUMAN

At the outset of this chapter, I identified three goals for this book: one has to do with messaging and counting; another, with what is special in the human experience of communication; and a third, with what kind of human is conjured by the countings and the measuring, on the one hand, and the claims that communications can allow a communion of souls, on the other. This last, I have argued, has led to the paradox of the age—our desire for communication and our complaints about its burdens. Underneath this paradox is a view about what human nature is and who we are.

Curiously, the literature on communication seems to be incomplete on this point. The literature suggests that human expressive artfulness can create something special and unique through the bonds of communication, and yet it seems to say nothing much about *who* is behind the communication. For example, I mentioned above Manuel Castells and his work on the network society. He has written numerous volumes on what this society is made of and how it is different from prior societies (in addition to those already mentioned, see Castell's 2002 book, *The Internet Galaxy: Reflections on the Internet, Business and Society*). For Castells, networks are superimposing themselves on other forms of human structure (such as hierarchical governments and business organizations). But humans scarcely appear in these networks, and when they do, they seem to be little more than information exchangers. It might be that his scope is too large: he tries to encompass everything and so characterizes nothing.

Perhaps other analysts might offer something better because their focus is narrower. Business people and management scientists ostensibly are more limited in their concerns than Castells is: his topic is society, and theirs is organizations. So many critics are commenting on the changes brought about by communication, especially communication enabled by the Internet, that they cannot all be listed here. As it happens, this community of researchers got to address the potential impact of the Internet and the associated explosion of communication technologies somewhat earlier than the sociologists. But nearly all offer a vision of humans as little more than information exchangers. See, for example, Philip Evans and Thomas S. Wurster's article "Strategy and the New Economics of Information" (1997), Evans and Wurster's book *Blown to Bits: How the New Economics of Information Transforms Strategy* (2000), and Robin Lissak and George Bailey's *A Thousand Tribes: How Technology Unites People in Great Companies* (2002). This particular line of reasoning reaches its apotheosis in Yochai Benkler's (2006) book *The Wealth of Networks*. In these (and many more), the human is simply a machine that produces, exchanges, and uses information.

Attempts to counter this distortion toward information and to bring a broader, more comprehensive view of the human aspect back into the discussion often seem to exaggerate by saying that organizations are about information exchange whereas communication between humans is about bodily things (passion, emotion, and touch). These researchers distinguish between organizational communication and private communication. But they are not the blog advocates who are concerned with bringing a personal flavor (even an angst) to organizational expression. the researchers I have in mind here want to make a bigger contrast—one between *disembodied* communication (via the Internet and also suited to organizational action) and *embodied* communication (the kind done when people are together and distinctly human). These researchers make contrast between words over distance and touch nearby, between the formal sterilities of language and the physical erotica of the body. See, for example, Barry Wellman and Caroline Haythornthwaite's *The Internet and Everyday Life* (2000), especially the chapter by Norman Nie, Sunshine Hillygus, and Lutz Erbring. See also Joseph Turow and Andrea L. Kavanaugh's *The Wired Homestead* (2003).[3] Again, these views seem to exaggerate one aspect of the human communicator at the expense of another and in so doing miss a whole range of potentially important dimensions and facets. Communication in organizations is sometimes about information exchange and sometimes about touch. Much domestic or personal communication is about bodily matters, but some is also about information exchange. What is in-between also needs to be accounted for and reflected on to understand what communication is and who makes it. Communication and the people who communicate are not to be encapsulated by these two opposites alone—fleshiness and wordiness.

Some commentators on the subject of the who in communication don't seem to have come from any particular discipline and hence might not get trapped in disciplinary prisms and concerns—like distinguishing organizational communication from other forms (such as Benkler) or elaborating the structural patterns of human relations that communications might enable (such as Castells), or defining those acts of communication that deliver human intimacies (such as Nie, Hillygus, and Erbring). Authors like Clay Shirky (see his 2008 book, *Here Comes Everybody: The Power of Organizing without Communications*) offer their own visions of the human that undertakes acts of communication. He explains that people communicate for three basic reasons—to share their knowledge (this is a kind of virtue in their nature), to display their vanity (this is less of a virtue and more a fact of life), and to seek conviviality (people cannot help being social animals, he claims). One cannot really doubt this simple ontology, but it fails to encompass the richness of what being a human might entail and how it might lead to various kinds of expressions or motivation (virtue, vanity, and conviviality notwithstanding).

Similarly, blog advocates who, like Shirky, don't seem to come from any particular discipline (like Scoble and Israel), seem to feel that expression via blogs is real whereas other forms of communication (especially by corporations) are somehow less real, more sterile, even insincere. Again, this may well be true in many cases, but it hardly helps us get a sense of who the person (or persons) behind the communication might be. These views don't help us properly understand what people might be seeking when they express beyond the most facile level. In their view, people communicate truly or not truly, with passion or dishonesty. This isn't enough to explain the intoxication of communication nor the complaints that intoxication engenders. Perhaps these visions seem inadequate because the modern technological landscape (the Internet, cellular phones, wearable computing) is so entrancing that authors who are free from any academic or scientific focus and those who are subject to disciplinary visions neglect to think about the human user (or subject) of these technologies.

The age when technologically mediated communication has reached its greatest volume might be encouraging commentators to forget the thing that is communicated (intentions) and the creature that creates and expresses these intentions (the human) and to focus instead on the sociological and organizational marvels that the technology enables. Castells is entranced by what Internet-enabled networks can do for social structures but not too interested in why people might want to say "Hello." Shirky is entranced by the fact of networked sociality and not by the complex nature of the humans who do the networking.

CONCLUSION

This temptation to overlook the humans who are doing the communicating is so powerful as to be almost ubiquitous. Take, for example, the view from what has come to be called *communications sciences* (or sometimes *media studies*). Central to this discipline is exploring the relationship between the human user (or recipient) of media content, especially broadcast content. When the discipline first emerged some twenty or thirty years ago, defining the media (and hence its message) was easy to do. But today, there are various sources of media, not just newspapers, radio, and television. The Internet has altered the landscape such that a plurality of channels now mediate content to the user. Communications science theorists have analyzed what consequences this change might have, and one of their conclusions is that people are becoming overloaded. This is the conclusion of W. Russell Neuman and colleagues' report *Tracking the Flow of Information into the Home* (2007), a study of media consumption in the United States from 1960 to 2005.

In this case, Neuman and his colleagues argue that a human can be treated as an information processor and that the processing they undertake is of words. Taking their cue from De Sola Pool's research in the 1980s (see Pool's 1983 article "Tracking the Flow of Information"), they argue that adults read 240 words per minute. With this base line, they analyze the time that the user has to consume the words sent to the home via the many channels that are "sent" or "pulled" into that setting. They conclude that there are too many words for the user to read or consume in the time available and that automated or intelligent systems will be necessary to select content on behalf of the user in the home of the future (11).

It strikes me as odd that a heterogeneous activity such as reading can be distilled into a simple metric like 240 words per minute. In this view, reading the back of a cornflake box is the same as reading a newspaper, a novel, a blog, a manual for a new washing machine—or a love letter. This view also makes the human choosing to do these different acts the same too. It makes reading a singular, mechanical act and makes the human equally mechanical.

This approach can be appealing because it allows a simple quantification, but it offers a rather feeble vision of the human that reads. As I noted with coauthor Abi Sellen in *Myth of the Paperless Office* (2003), reading is an activity that is easy to oversimplify, and *reading* is a catch-all phrase for a number of activities that reflect something of the human in question—who they are and what they are seeking to do when they read. As it happens, only some activities labeled "reading in the workplace" can sensibly be understood in terms of speed. Indeed, speed is not the important dimension to be applied when thinking about reading technologies for work, for example.

This is also likely to be the case in the home setting, the one that Neuman et al. concern themselves with. As Alex Taylor and I noted in 2003 (115–126) (in a study about television consumption), when people go home and pick up a newspaper or switch on the TV, they are not approaching that action as merely an information processing task. They might be doing so simply to turn themselves off. Reading the paper and watching TV here are ways to end the day's work and begin the day's leisure. These activities are not to be understood as being done on the basis of a choice between content formats or types or in terms of speed. However many words are read or news items watched, this type of activity is concerned with using twenty minutes to make a transition between work and home. And this, in turn, says something about the kind of person who chooses to break up their day in this fashion (not all people will do so, after all).

In offering quantitative measures of an activity such as reading (and media consumption more generally), Neuman et al. are not being disingenuous—even if they are being a bit lax in their science. For their countings of media input and media consumption are typical not just of their discipline, but of other disciplines too—sociology, organizational and management science; it is typical of the popular commentators on digital connectivity, such as Shirky. But it is also typical of the ways that most people tend to think of themselves today—and not just those who think of themselves as knowledge workers. I am thinking of all of us when we take off our work hats, go home, and orient to our lives in ordinary commonsense ways. We *do* look at the infinite number of channels on our TVs and wonder how we might consume them all. We *do* look at the news on the Web and wonder how much time we could allocate to reading it all. We *do*, beyond this, start looking at ourselves in terms of inputs and outputs and start treating our communicative habits and our mediated communications as visible measures of *overload*. Hence, we notice these objects above all else. We look at the numbers of messages received and wonder how we can balance the delight we get from their receipt against the labor we need to put in to reply. As we do so, we naturally turn to measures of our time and the pressures on it since this seems the most precious resource of all. We start from the *assumption* that quantitatively demonstrable overload is the measure of our age, and so we look at ourselves and our activities with that in mind and make it so. If we don't start from this point, we soon learn that we ought to by the narratives produced by the experts—the media specialists like Neuman et al. We thus find ourselves ignoring the fact that when we read the back of a cornflake box at breakfast, our eyes are simply caressing the words and not consuming them and when we switch on our home computers and gaze at the evening news on our Web feeds, we aren't digesting what we see but are waiting for our minds to unravel the news in our *own* affairs, not in the world at large. In both cases, our bodies have been consuming words but not in the sense that Neuman et al. mean it and not in the sense that we lazily apply ourselves when we look up at the clock after these events and wonder, "Where has all the time gone?"

As Marta Banta notes in her book *Taylored Lives: Narrative Productions in the Age of Taylor, Veblen, and Ford* (1993), we seem to have become transfixed by these ways of thinking about our endeavors. But as we do so, so we disregard other ways of thinking about what we do. Banta's analysis was written before the onset of this concern with communications overload (it was about the desire to measure and monitor every activity to manage ourselves better). But questions about why people communicate and who the communicating human might be are as old as philosophy itself, perhaps even as old as language. Perhaps the best history of thinking about this subject is John Durham Peters's *Speaking into the Air: A History of the Idea of Communication* (1999). Peters is particularly good at exploring the conceptual implications that various technologies have on the structure or hopes that are embedded in what he calls the "metaphysics of the idea" of communication. New technologies alter this metaphysics, he shows. For example, the invention of recording devices in the nineteenth century that could copy and replay human voices helped cultivate the idea that people had a speaking soul that was trapped inside a body. The hearer of these early recordings of people speaking thought that the real person was speaking in a ghostlike manner. This lead to an emphasis on "innerness"—on a thing (a spirit, perhaps, or the talking soul) trying to get out and transcend the body and its "skin." This evoked a greater sense of the subjective—of how solipsism was solved through communication. Even some of these words (such as *solipsism*) were constructed as a result of such metaphysics, he says.

Peters goes on to say that there is a contemporary metaphysics too. It seems to resonate with what I have said about some claims about human communication—the blog advocates who suggest that human passion is coming back into organizational exchange and others who say that some visions of communication beyond the contexts of organizations lack a concern for the body that is so vital to human connection. Peters gets to a similar view but from another direction. He says that our attempts (in the late twentieth century) to devise ways of seeing each other via video and our attempts to offer more sensual aspects to communication to augment sight (like touch) draw attention to what he calls the erotic aspect in the act of communication. His view is not that people have always communicated for erotic reasons but that the late twentieth century and early twenty-first century have led us to think and act as if being in touch means just that—something to do with the body. Our technologies of communication have helped create what we think we are and hence give motive to our acts of communication.

One can imagine how contemporary metaphysics are being built given our twenty-first-century technological landscape. But Peters doesn't tell me enough about how the world we live in has come about, nor does his work seem to recognize that there are many ways of orienting to the world, each of which might have its own metaphysics. I am not sure that the erotic motivated the design of the first video phone, for example, even if the video phone might have highlighted the erotic once assembled. I would like to look at what the designers were thinking and what the first users thought too. In Peters, one confronts the trouble and the pleasures of *cultural theory*. The pleasures are in the delightful drawing of links and allusions and the making of metaphors and contrasts. But the trouble is that one cannot tell where the contrasts and metaphors end and real evidence starts. Above all, one cannot grasp what people might be doing in the landscapes he describes. He offers possibilities, but one's instinct remains doubtful. I am not sure that, in the nineteenth century, listening to recordings of voices made people think differently about communication in just the way that he says, just as I am not sure that a video call emphasizes the body in just the way that he says—erotically.

Somehow, we need to be both more historically detailed and more willing to accept that the world we have come to assemble might not be best thought of as a nice tidy system—a cultural framework. So what might it be? How does one account for the landscape around us? What vision of the human have we been using in our affairs to make the world as it is? It is to that we now turn.

NOTES

1. Envelopes were first used in about 1840.

2. Similarly, the claim that the Internet is good for you or bad for you—the so-called Internet paradox—is a debate that is built on shallow premises. As James Katz and Ronald Rice note in their thoroughly empirical book *Social Consequences of Internet Use: Access, Involvement, and Interaction* (2002), having evidence at hand does not ensure that claims are evidence based.

3. Turow and Kavanaugh's book also contains Robert Kraut and colleagues' famous 2003 recanting of their prior claim that use of the Internet makes people less social, something they argued in the 1990s. After a decade, they were no longer certain about this point.

REFERENCES

Austin, J. L. 1962. *How to Do Things with Words*. Oxford: Oxford University Press.

Banta, M. 1993. *Taylored Lives: Narrative Productions in the Age of Taylor, Veblen and Ford*. Chicago: University of Chicago Press.

Benkler, Y. 2006. *The Wealth of Networks: How Social Production Transforms Markets and Freedom*. New Haven: Yale University Press.

Boyd, D. 2008. Taken out of context: American teen sociality in networked publics. PhD diss., University of California, Berkeley.

Castells, M. 1996. *The Rise of the Network Society*. Oxford: Blackwell.

Castells, M. 2002. *The Internet Galaxy: Reflections on the Internet, Business and Society*. Oxford: Oxford University Press.

Castells, M. 2009. *Communication Power*. Oxford: Oxford University Press.

Cavanagh, C. 2003. *Managing Your Email: Thinking outside the Box*. Hoboken, NJ: Wiley.

Derrida, J. [1972] 1988. Signature event context. In G. Graff, ed., *Limited Inc.* (1–23). Evanston: Northwestern University Press.

De Sola Pool, I. 1983. Tracking the flow of information. *Science* 211:609–613.

Dutton, W., and E. J. Helsper. 2007. The Internet in Britain: 2007. Oxford Internet Institute, University of Oxford. Accessed October 3, 2007, from http://www.oii.ox.ac.uk/microsites/oxis/publications.cfm.

Evans, P., and T. S. Wurster. 1997. Strategy and the new economics of information. *Harvard Business Review* (September–October): 71–82.

Evans, P., and T. W. Wurster. 2000. *Blown to Bits: How the New Economics of Information Transforms Strategy*. Boston: Harvard Business School Press.

Fisher, C. 1992. *America Calling: A Social History of the Telephone to 1940*. Berkeley: University of California Press.

Gershuny, J. 2000. *Changing Times: Work and Leisure in Postindustrial Society*. Oxford: Oxford University Press.

Gershuny, J. 2007. Conclusion: A slow start. In B. Anderson, M. Brybin, J. Gershuny, and Y. Ragan, eds., *Information and Communication Technologies in Society: E-living in a Digital Europe* (274–280). London: Routledge.

Hamill, L. 2008. *Time to Communicate. A Technical Report for MSR*. Cambridge: Hamill & Hamill Ltd and Microsoft Research.

Hamill, L. 2010. Communications, travel, and social networks since 1840: A study in agent-based models. PhD diss., University of Surrey, Guildford, England.

Henkin, D. M. 2006. *The Postal Age: The Emergence of Modern Communications in Nineteenth-Century America.* Chicago: University of Chicago Press.

Hewitt, H. 2005. *Blog: Understanding the Information Reformation That Is Changing Your World.* Nashville: Nelson Books.

Katz, J., and R. Rice. 2002. *Social Consequences of Internet Use: Access, Involvement, and Interaction.* Cambridge, MA: MIT Press.

Kraut, R., S. Kiesler, B. Boneva, J. Cummings, V. Helgeson, and A. Crawford. 2003. Internet paradox revisited. In J. Turow and A. L. Kavanaugh, eds. *The Wired Homestead: An MIT Press Sourcebook on the Internet and Family* (347–384). Cambridge, MA: MIT Press.

Lenhart, A., M. Madden, A. Rankin, and A. Smith. 2007. *Teens and Social Media.* Pew Internet and American Life Project. Available at http://www.pewInternet.org.

Lissak, R., and G. Bailey. 2002. *A Thousand Tribes: How Technology Unites People in Great Companies.* New York: Wiley.

McRobbie, A. 2000. *Feminism and Youth Culture.* London: Routledge.

Neuman, W., J. Park, and E. L. Panek. 2007. Tracking the flow of information into the home: An empirical assessment of the digital revolution in the U.S. from 1960–2005. Available at http://www.wrneuman.com/Flow_of_Information.pdf.

Nie, N., D. S. Hillygus, and L. Erbring. 2000. Internet use, interpersonal relations and sociality. In B. Wellman and C. Haythornwaite, eds., *The Internet and Everyday Life* (215–245). Oxford: Blackwell.

Office for National Statistics, 2005. *The Time Use Survey 2005.* London: HMSO.

Peters, J. D. 1999. *Speaking into the Air: A History of the Idea of Communication.* Chicago: University of Chicago Press.

Rodzilla, J., ed. 2002. *We've Got Blog: How Blogs Are Changing Our Culture*. Cambridge: Perseus.

Scoble, R., and S. Israel. 2006. *Naked Conversations: How Blogs Are Changing the Way Businesses Talk with Customers*. Hoboken: Wiley.

Sellen, A., and R. Harper. 2003. *Myth of the Paperless Office*. Cambridge, MA: MIT Press.

Shipley, D., and W. Schwalbe. 2007. *Send: The How, Why, When and When Not of Email*. Edinburgh: Canongate Books.

Shirky, C. 2008. *Here Comes Everybody: The Power of Organizing without Communications*. London: Allen Lane.

Sproull, L., and S. Kiesler. 1992. *Connections: New Ways of Working in the Networked Organization*. Cambridge, MA: MIT Press.

Taylor, A., and R. Harper. 2003. Switching on to switch off. In R. Harper, ed., *Inside the Smart Home* (115–126). Godalming, UK: Springer.

Turow, J., and A. L. Kavanaugh, eds. 2003. *The Wired Homestead: An MIT Press Sourcebook on the Internet and the Family*. Cambridge, MA: MIT Press.

Urry, J. 2007. *Mobilities*. Cambridge, MA: Polity Press.

Wellman, B., and C. Haythornthwaite, eds. 2000. *The Internet and Everyday Life*. Oxford: Blackwell.

3 ABSENCE TO PRESENCE

PREAMBLE

In 1904, Edouard Estaunie coined the term *telecommunication* by merging the Latin word *communico* (impart or share) with the Greek word *tele* (distance). He had in mind a word for any technology that used electronic signals to exchange information. We still use the term in pretty much the same way one hundred years later, although in a more encompassing fashion. The historian of telecommunication, Anton Huurdeman (2003), uses it as a label for any "technology of information transport." Paper mail and motorcycle couriers would presumably fit into this category. We no longer think of *telecommunication*, however, as the word for our communications-obsessed times. Instead, the words *computer* and *Internet* have become synonyms for the all human desire to be in touch with others. Here I am thinking of not the evolution of words but simply the use of everyday phrases like "Are you on the Internet?" and "I am much better at keeping in touch with a PC." These phrases convey that some thing (a PC, a mobile phone, the Internet) is like an umbilical cord linking each of us in principle—an umbilical cord between ourselves, our friends, and the world at large.

In this chapter, I try to define the kinds of human vision (that of the user who desires to be in touch) that underscore the inventive landscape that I see around me. Key to this vision has been the idea that more is better and that we should keep our inventive imagination to the grindstone to produce more ways of being in touch. This approach has produced a landscape in which we talk about being in touch via computers as if it were commonplace and also expect that this landscape will keep changing as new tools for richer, more, and hence better communication appear. In later chapters, I ask whether we need more communications channels, but here I try to answer why we keep inventing them. My underlying claim is this inventiveness is driven by an idea about what is human.

For twenty years, I have worked in places that have been concerned with inventing the technologies that are of interest to us now. At first glance, this world might not seem to have been driving the communications landscape around us. I currently work at Microsoft Research and have also worked at Xerox PARC's European lab, EuroPARC. One might imagine that the research carried out in these places was about work life and not about being in touch in private life. But in practice, these landscapes were being constructed by people with a passion for computer-mediated communication. This passion has been bound up with the emergence of the Web and the protocols that allow data traffic to move around the world at enormous speed and in huge volumes.[1]

In this chapter, I characterize a model of the interacting human that my colleagues and I had in mind when we invented in these workplaces. This model has two elements. The first emphasizes the bodily aspects of communication and hence treats physical distance as the problem that technologies of communication solve. This view generates lots of interesting ideas as well as new technologies for communication, many of which we see around us now. But this vision of the human also misses out important aspects of the *communication act*. As we saw in the previous chapter, letter writing might have started as a way of solving the problems of physical separation, but once cultivated, letter writing has an altogether different effect on those involved. It doesn't bring them together in a physical sense but in a moral one, creating a new sensibility for human bonds, bonds mediated in this case by pen, ink and the transporting envelope.

The second view emphasizes the processing limits of the human. This treats the human not just as a body located in space (and therefore separated from other bodies through space) but as a machine that processes information, including acts of communication. Words are one of the substances that this processing machine deals with. In this view, there are objective thresholds beyond which people can no longer process information. If the body emphasizing view has predominated in the past, this processing view is now central to what motivates much of the research that is being conducted today on *communications overload*. This research looks at such things as how to tame interruption and communication excess. If human communication is merely a form of attention giving, then attention thresholds can be used to determine what to invent for people so that they don't communicate too much.

I am not sure that this is the best way of thinking about human communication, although as with the body emphasizing view, it provides something to work on for those who are in the inventing business. This view of the human may be appealing for pragmatic purposes, but it takes the expressiveness out of communication and recasts communication as a measure of volume. One problem that my colleagues and I deal with is the shift between our professionally pragmatic notions that there are such limits (that human expression is a measurable entity that has a volume) and our ordinary, commonsense notions (that human expression is rich in meaning and inflection and that quantity is only occasionally the appropriate measure to apply when seeking to describe, account for, or assess the expressions in question). When deploying common sense, everyday reasoning, we might use the word *quantity* when we are thinking of moral overtones ("You said too much," for example), whereas in our professional lives, we use the word *quantity* when we are thinking of what a body can process—what our ears can hear and our eyes can see. These are not the same.

INTERACTING INDIVIDUALS

Some twenty years ago, my colleagues and I were toying with two or three systems that let us undertake certain kinds of communication acts, and today I still see colleagues playing with variants of some of these concepts and inventions. What motivated us then had to do with a mix of hopes and expectations. But beyond this, we were motivated by notions of what a human is, what an interacting human does, and what our inventiveness ought to enable. We are entranced by the same view today.

In the early 1990s, I was fortunate enough to work at Xerox EuroPARC in Cambridge. For many reasons, EuroPARC was a curious place to work, and one was the predilection of many of its staff for video-mediated communication (VMC). Because of this, the entire EuroPARC building was wired for sight and sound. Nearly every office had a coaxial cable that linked it to every other office and that conveyed high-quality audio-video connections between offices. These cables and connections could create what was called a *media space* between any two or more offices (see William Gaver et al. 1992). Part of the fun of working at EuroPARC was the shock and glee that was provoked by the media space. Some journalists were appalled and called it Big Brotherish. Others were enamored at how the media space pointed toward the prospect of bringing together people who were separated by great distances.[2]

Many projects investigated the utility of the set of systems that comprised the computational infrastructure of EuroPARC's media space. Looking back, a couple of these systems are notable. Shared editors—what were most often were called *shared whiteboards*—allowed people in different offices to type text into a document at the same time as someone else was typing in another office. One of us would type text into a window on a screen (the *editor window* of the application), and our colleagues in another room saw this being typed into their view on the editor window and could type on the next line of that window. This would be visible to the person who created the first line. The system did not indicate who wrote what. But typically, one could tell simply by the turn taking: one person wrote, a second wrote, the first replied, and so on. That was the gist of it.

These editors were being investigated for their use in what was called *knowledge work*. Knowledge work might be a vague term, but we were in interested in supporting it, especially when the participants were separated by distance, because new organizational structures were increasing the likelihood that such work would be undertaken by individuals spread around the world. As organizations became increasingly global, knowledge work was being done in New York, London, and New Delhi, and technologies were required that could support such distributed activities. It was thought that the distance between the individual offices in our building could be treated, analytically, as proxies for bigger ones in the real world.

Nothing much came of the shared whiteboards, whatever the merits of investigating knowledge work. We did not manage to implement them in any effective way in real work outside of our lab, nor did we use them much for our work. Although these tools were central to what our managers stated was our research goal (or topic) we did not find them hugely useful for our own activities, our own knowledge work, nor did we find then particularly interesting. They seemed too prosaic to be worth investigating, although some visiting researchers found interesting aspects to them, which they published in the groupware and computer-supported collaborative work (CSCW) literature. Our managers remained vaguely hopeful that we might evolve these shared whiteboards in novel ways, bringing a Xerox sheen to what was a fairly crude technology. But we did not. Other technologies, especially the audiovisual aspects of the media space, enticed us away from the whiteboards.

Nevertheless, we used these shared whiteboards for ourselves—not for knowledge work (to help us write reports or share profound ideas) but for play. They became devices for laughter and mischief, and using them could be fun. We used them for jocular asides about the burden of deadlines, and occasionally to make plans about after-work activities. At least one romance at EuroPARC flourished with this form of exchange. What we were playing with then was the genesis (in a convoluted and indirect way) of what is now known as *instant messaging*. Although our managers thought that these tools would be used by professionals in their knowledge practices and my research colleagues and I assumed that these tools would be work-orientated applications, some twenty years later one can see that they have been crafted into tools for sociality. These knowledge work technologies have ended up allowing, for example, teenagers to keep in touch with each other. Instant messaging is a way of communicating not for work's sake but for friendship's sake. Tools for serious things like work have ended up being used as tools for other things, for friendship and for play in social life (see Harper 2005a and Harper and Hamill 2005).

It now seems perplexing that we believed that these applications were to be thought of only in terms of work. As I say we did not use them to support our own activities even though we were professionals doing knowledge work of a kind. We found fun and utility in them through their playfulness and in supporting not professional engagement but our social selves. We were doing then pretty much what teenagers do now—indulging in what the French sociologist Michel Maffesoli suggests, metaphorically, in his *The Time of the Tribes: The Decline of Individualism in Mass Society* (1996) is a contemporary form of Dionysian tribalism. He has in mind the desire of our "modern selves" to celebrate the temporary, daily, bodily togetherness of work and social tribes, whichever tribe that might be—research lab tribes included. I think there is certainly something in this metaphor.[3]

Be that as it may, there is a bigger question that is brought to mind by this. The fate of this technology can lead one to ask whether there is something peculiar in the relationship between those who do the inventing and those for whom the inventing is said to be done. Are inventors justifying their ideas on the grounds that someone (perhaps not themselves) will use their inventions, and are these ordinary people in the real world something of an artificial construct? One might say that one of the ways that we work in my world is by making users "other" in various ways and that, for us, this otherness was (and continues to be) simply other than ourselves.[4]

Over the years, I have heard my colleagues justify some ideas with the assertion that real people will use them, even if the colleagues in question (the inventors of the thing) don't say who these people might be. We all can drift into fatigue or laziness and might not really have in mind a particular real-world user. Sometimes, phrases like *ordinary people*, *real people*, and *people in the real world* get used to justify technologies that my colleagues and I haven't really figured out the purposes of for ourselves. We think we have devised something that is of use, but for what and for whom are beyond our comprehension. Ordinary people have lots of faces and lots of desires, and so we use this label as an expression of hope that some one, somewhere, for some reason or other will indeed like our contraptions.

In some instances, we don't have a clue about who might use our inventions, but this is not common. My colleagues and I are pretty thorough in our research and most often have a user in mind, but it might not be a view of the user that is held outside my kind of workplace. This view might not always be well articulated, but nevertheless it is held by most of my colleagues. They held it then, and they hold it now. The example of shared editors hints at what this view is. My description of another technological experience that we shared at the EuroPARC media space will explain a great deal about what kind of user we had in mind and how this vision motivates us today.

WAYS OF LOOKING IN MY WORLD

While we were using these shared editors, another project was adding some refinements to the media space, especially to its audio-video conferencing capabilities. When I first arrived at EuroPARC, the media-space system was fairly simple. Each office had a high-quality camera that was mounted on a wall or tripod or attached to shelves and that allowed others who used the media space to see you. Speakers were fitted in the room wherever space could be found, and a monitor allowed a user to see the other office when a connection was made. All these were separate from the workstations that we were using, although we used the workstation to initiate a connection. This arrangement meant that these devices usually were separated from one another. If one looked into the camera, then one might well be looking away from the monitor on which one could see a colleague. The same would hold true for colleagues. So when one made a connection to another person's computer, one would look at this person on one's own monitor, but the other person would appear to be looking elsewhere—not at you. They too were looking at their monitor, wondering why we were looking away. In both cases, the camera would be far enough away from the monitor for this *incongruency of perspective* (as my colleagues expressed it) to occur. Some of us found this vexing or amusing. "Look at me when I am talking!" someone would shout. "I am!" someone else would reply. Some of my colleagues, however, were convinced that this problem needed fixing. They devised solutions—inventions of sorts—that solved the problem. These solutions were not difficult. Somehow the cameras needed to look at the subjects as if the cameras were looking from the same point of view as the monitor. Both the looking and the glancing needed to be the same.

In the early 1990s, the only way to achieve this fix was to assemble and box up mirrors (like a periscope) around the monitors so that the line of sight of the camera appeared to be hidden within the monitor. This improvement to the system was called *video tunnels*. Various experiments were undertaken to investigate whether users (us) felt at ease with the system. One particular concern was whether users could better understand what others in the system were seeing. This system allowed us to know that when a colleague looked away, he wasn't looking at our own picture, for example, but instead might have an object (a document, say) that he wanted us to view. There might be a *mutual referent*, as it was grandly put. Efforts were made to write up these findings for publication, but these faltered; and within weeks after the experiments were completed, the system was packed up. The periscope-type boxing was large and unsightly, and no one seemed to think that video tunnels made a significant improvement. The media space returned to its former state.[5]

I certainly found it easier to use a system when the eyes of a fellow participant seemed to be looking at me. But much of my own enjoyment with the media space had nothing to do with the ability to look into another's eyes as he or she gazed into my own. It was fun merely to peer into another room to see what was happening there. Sometimes we would use the media space to call out to each other between rooms separated by several floors and see if anyone was around. This was especially so between reception (the ground floor) and the administrative offices (the top floors). This play did seem to be a kind of Dionysian tribalism, as Maffesoli would have it—all the more so as I look back now and wonder at how much I played when I was younger.

I would like to continue to examine the view of the user that motivated this enhancement of the media space by recounting how years later, my colleagues in another research lab in the same city are endeavoring to invent something very similar. But this time, they are devising something without the ugly boxing, the heavy camera, or the coaxial cables. We once had to make do with very large cameras, and now video cameras can be very small indeed. At that time, the best we could do was stream the video data over the network, and now we can process it and interrogate it. This makes a considerable difference to what can be assembled or invented.

In the new system that I have in mind, each monitor in the link (a link between two offices, say) has a small stereo video camera attached to its top (this is actually two cameras placed side by side). As before, the cameras look while the monitors display what is to be seen. But in the new system, data from the cameras are processed with so-called intelligent algorithms so that the system can recognize and follow the movements of those looking at the monitor. In simple terms, the cameras produce stereoscopic data that the computer can analyze to identify where the user's head is. There are various ways of doing this, such as tracking movements of the head (as a three-dimensional outline) or tracking specific points on the head (such as eyes, mouth, and tip of the nose). Irrespective of the tools (or *keys*) that the system uses, after the tracking is processed, the computer can adjust the image shown on the remote monitor to ensure that when the user is gazing at his or her own monitor, it looks as if they are gazing through the cameras at the remote viewer. The actual distance between the cameras and the screen is dissolved. The result of this system is delightful, albeit slightly odd. The remote person's eyes are not really gazing at you; they are looking at a screen displaying a virtual you. But it affords a playful paradox: it's nice to be looked at, but after a while one tries to see how far one has to go before one cannot be looked at—to go beyond the processing of the system so that it cannot correct.[6]

What was the goal behind these two systems? In one way, it seems obvious and perfectly reasonable. During a video conference (or videocall), it would be nice if the person you were conferencing with appeared to be looking at you rather than at something else. Both systems provided a feeling of greater politeness and social grace. The point I am making, though, has to do with the relationship between my colleagues' ideas of what to design for (ideas related to what the human actor or user of the system is) and how to deliver, through design, this sense of grace. When together or when using a conferencing system, people somehow manage to coordinate their gazes and orient their behaviors to produce for each other a sense that they are in touch, of a mind, and doing something collaboratively. There is a great deal of empirical literature on how this sense of jointness is achieved, especially in the ethnomethodological canon (see, for example, David Sudnow's 1972 book, *Studies in Social Interaction*). There is a similarly large literature reporting on the moral dimensions of looking and glancing. Researchers like Egon Bittner have explored how *not* responding to a look can be seen as social rebuff, for example, just like not answering a hello. His 1977 paper "Must We Say What We Mean?" points out that meaning is conveyed even when we don't speak or look. Our actions embody our intentions and are thus visible for others to see. But that this is so was not what motivated these systems. What underscored them was a much simpler vision of human interaction or interacting humans that emphasized just the body mechanics of the interaction and lost sight of this moral dimension and the ways that people create a sense of joint endeavor when they communicate. This approach splits the human into a body and mind and designs only for the body while treating the mind (and all it might stand for, such as mindfulness and intentionality) at best whimsically—as something that will be satisfied indirectly, if at all.

FITTING TOUCHES

Why is this significant? For one reason, I am trying to get to the view of the human that motivates us, and these examples are of systems developed both some years ago and recently. These technologies are not isolated attempts to build "solutions" that are determined by a particular view of the human. My colleagues have been developing another conference system of sorts that has a similar concept of body mechanics as the key of human communication behind it. This view is persistent and common, in other words. The systems I have mentioned thus far have concerned themselves with the how of looking at faces, and the system I now describe concerns itself with the touching and moving of hands.

The system in question is again a video conferencing system of sorts. The concern is still to solve the problem of knowledge work over distance. This system uses a stereo camera and some clever object-recognition software to notice, for example, when hands are placed on a digital document, when they point, or even when they appear to erase a word. The inventiveness involves making such movements recognizable to the system. First, the machine must be made able to discern shapes (in this case, the shape of a hand and a finger) from other shapes and forms that the cameras might see. Second, this ability is combined with the capacity to determine movements of these things when the movement in question indicates something. Hence, a finger's movement can be seen as a gesture, for example, not a mere shadow moving over a document. The result is a system where users can get to grip their digital documents. They can move their hands over a document to mark, edit, remove, and paste words and images in the document. The movements of their hands and the gestures of their fingers are converted by the machine into the equivalents of mouse pointings and clickings. Such a technique is not devised to replace mouse pointing. Rather, it is a method that allows people to interact with documents over distance. One person points and shakes his finger over a document; a person somewhere else can see this interaction and can have the consequence of this interaction made manifest in the document they are working on. A paragraph can be highlighted or moved, and a picture repositioned. This is a hand interacting with a document that is digital and that is shared (see Shahram Izadi and colleagues, "C-Slate: Exploring Remote Collaboration on Horizontal Multi-touch Surfaces," 2007).

With this and the prior examples, one can begin to see what my colleagues have in mind about how to conceive of the user. The idea is that, whatever people want to do, they can be aided in that task by systems that replicate what one might call the *physical geography* of the interactions in question. In the first two examples, we imagined that if we were designing a communication experience for people who want to communicate to others far away, then one way of supporting that would be to invent a system that would let the people in question glance and gaze at each other when they are remote just as they might when they were together. What we needed to design for was what might be called the *interactional geography* of lookings. In the example of the more recent system, we were also concerned with remote communication but in this case in interaction and communication about documents. Here a system was invented that meant that how someone uses his or her hands when dealing with a document could be conveyed to someone remotely. Here the *geography of touch* mattered.

Calling these things geographies of touch and looking might make this vision of human needs in communication technologies seem obscure, even arcane. But my description is designed to make strange what is in fact commonplace in the world of inventing computer-mediated communications systems. Making it strange might help us think about it. What motivates a great deal of the inventiveness that I see around me is a concern with the human body, with the bodily mechanics of human interaction with computers, and with hence with communication through and with computers to other people. Inventiveness in my world is about devising ways of allowing the mechanics of the communicating individual to fit together. In this vision, what is said in communication, why something might be said, and what may be the consequences of the saying of something don't matter. The how of the saying—the manifest behavior entailed when that thing is said—is what counts.

In this vision, what is exchanged in a communication might be glancings, pointings, gestures, as well as the written or spoken word. In all cases, my colleagues and I seek ways of replicating the body mechanics of those doings—so that they are conveyed over distance and so that the mechanics (the sounds, the glancings, the pointings) may be conveyed as an assembly. The goal is to replicate the interaction of multiple, communicating bodies. Glancings need to be synchronized, listenings need to be two way, and distractions ought to be shared if they are to be oriented to by another body's lookings and glancing.

BEYOND MY WORLD

Different research labs will naturally have their own preferences and ways of doing things. It could be that both EuroPARC and my current establishment have similar views. Hence both focus on the interlocking of bodies. But I now want to point out that another giant of the IT world has been producing a technology of communication that builds on a similar vision. Google has recently launched an application called Wave. Google staff explain that this offers a rich communications experience. With Wave, users can exchange words via an instant messaging type application, share documents and presentations and, if they so wish, connect via video. Wave offers multiple dimensions of being in touch.

Wave does this in a way which is similar to the application and devices mentioned above. Not technologically, but in terms of what is thought to be the essential properties of human communication that it satisfies. If the above applications allow an interlocking of glances and touches, of pointings and mutual references, then Wave also turns around a model of the communicating human, but its design highlights another feature of this vision.

Wave turns around the assumption that communication is best done when it is in real time. Bodies don't want to interlock with ghosts of bodies that have long since departed; fingers don't want to point at things that others won't see until some later time. The designers of Wave assume that humans want to communicate in the here and now. In this view, a written letter is a poor attempt at getting bodies together—in "real time," in the present. Indeed, if one looks at the Google presentation of Wave on the Web (there is still no scientific research reporting on the use of the application), this is precisely what the inventors say: letters are a technologically archaic and poor surrogate for the real human need, which is to be together, communicating without delays caused by sending signals (messages) across distance.[7] If people could use technology that could deliver this sense of being together then they would not write letters. Letters enable only asynchronous expression. One person creates their communication and sends it; the other waits until it arrives and sends their reply later. Letters are an abomination of dithering in this view.

It is not only the interlocking of bodies that communication technologies should manage, then, it is also the interlocking of those bodies in specific moments of time. If Xerox and Microsoft want to bring the mechanics of communicating human bodies into a mutual assembly, then Google wants to bring them together in a way that distance does not create time delays. All communication should deliver what in contemporary parlance is called presence—a sense of physical togetherness in the here and now. This is what good design with digital technologies can provide.

A VISION OF THE HUMAN USER

Numerous questions follow from this. Why this view? Why not another? Will bodily communicative acts ever be entirely replicated remotely? Will the problems of time delay of signals sent over distance ever completely disappear? On both counts this seems unlikely, but what difficulties arise here? The examples that I have used seem rather limited and essentially have to do with knowledge work and conferencing of various kinds. How should we view the daily assault of messages, callings, and blog postings that I mentioned at the start of the book? These don't seem like time framed body mechanics, so where have they come from? What inventive landscape produced them?

Some of these questions are certainly more important than others to our concerns. The question of where this view of the human comes from is I think the most important. It appears that my colleagues and I (as well as those in similar technology businesses) have a vision of the human (the user) and of the actions of the human when they communicate that is not the one used in the everyday world. Ordinarily people don't think of fitting their bodies together when they communicate (or rather, sometimes they do but not often); nor do they think that being in touch is *always* about being together in real time, in the here and now. Our ordinary selves have been brought up with the idea that communication is sometimes about this, being together and sharing a place and time, but it is also an art, and when done with finesse, creates a sense of connectivity that is beyond time and space.

In other words, although sometimes my colleagues and I use the phrase ordinary user, we don't have in mind a comprehensive vision of people communicating; we reduce their communication acts to something very particular. Following Durham Peters's formulation (mentioned in chapter 2), I think that my colleagues have a metaphysics—a set of ideas about what the human communicator is. In the examples above, it seems to me that important aspects of these ideas in question derive from various sources within computer science, the most influential being a founding father of the field—Alan Turing. He encouraged a view that splits the human into either body or mind and tends to disregard the latter. This might not be a precise representation of Turing himself or of the detailed claims he made in his various papers (there is no mention of mind body dualism in his work, for example), but it is a fair comment on the consequences of his view on the inventive processes that are concerned with technologies for human communication. These processes are lead, in the most part, by people trained in computer science, and even if they are not trained in it, the metaphysics of this view certainly seems to predominate. If my own experience is anything to go by, then taking this Turing theoretic view is the starting point of the enquiries that have lead to the communication systems I have described.[8]

In the various papers on intelligence and the computer machine that he wrote in the 1930s, 1940s and 1950s, Turing claimed that one may understand intelligence by treating it as solely to be measured in external performance—in behavior (these discussions are encompassed in his work between the publication of his 1936 paper "On Computable Numbers, with an Application to the Entscheindungsproblem," loosely meaning the "decision problem" and end, more or less with his 1950 paper "Computing Machinery and Intelligence"). Whether a machine or a person performs the behavior is irrelevant, he proposed. This had methodological advantages for him, as he saw it, because it offered a route around the problem of moral judgment. A person could not be relied on to describe a person's action as intelligent or otherwise for fear of upsetting the person who was being judged. Thus, in Turing's perspective, *only behavior* must count, and the source of that behavior must be hidden from view. Turing did not seem to be worried about reducing the concept of intelligence from something that is rich, complex, and subtle, that is used in diverse ways to achieve different sorts of understandings, purposes, and descriptive values in ordinary language and life. He preferred making the concept a mere label for a binary opposition where an action is either intelligent or not (for a review, see Piccinini 2003; see also Shanker 1998). In so doing, a particular consequence ensued. Those who adopt Turing's view assume that what goes on inside the machine itself is not only invisible but also somehow tricky and best avoided. Computer scientists ought to steer clear of the moral implications of judgment or the moral aspects of a mind's thoughts—if they are to do science as Turing conceived of it.

This hasn't meant that computers scientists subsequently haven't looked at what goes on inside the head and attempted to do a kind of science. But they have done so from the presumption that one starts with bodily behavior and then has to construct, with external evidence, ideas about what goes on inside the head. For some of those who have taken this Turing theoretic view seriously, if a computer machine can use a program in some act, some body performance (like a physical move that a player makes in a game of chess) and if this is equal to the body movement of a human, then perhaps the human mind is computerlike. In their opinion, research should be undertaken to discover the code that makes the mind work. Much of contemporary cognitive science, for example, is a product of the possibility that the human can be divided into body and mind and assumes also that the mind is a processing machine, just like a computer. It is a view commonly held elsewhere too, such as in neuroscience. David Marr, for example, argues in *Vision: A Computational Investigation into the Human Representation and Processing of Visual Information* (1983) that a computer machine ought to be able recognize a visual object (via various forms of Bayesian prediction, for example) in the way a human mind processes and hence sees (for a review of this and various other ways Turing theoretic computational dualism has become ubiquitous, see Sheutz 2002).

Turing has been enormously influential in many ways, and his view has its roots in earlier ideas—for example, Cartesian dualism and clock mechanisms, which once provided inspiration for understanding the human body (see Hacker 2007, 233–257). But here I am recounting some of the inventiveness I have seen around me in terms of a certain vision of the human. A concern with body mechanics in particular points of time and space has motivated us—not to investigate philosophical ideas about humanness but to pursue pragmatic goals. We have sought visions of human endeavors and human communicative actions that could lead to new ideas and inventions. A body-emphasizing, time fixed view has helped us to achieve this. It has highlighted problems and needs as we saw them—the kinds of things we thought we could solve.

This view ought to be judged by what it allowed us to do—to produce new inventions, new designs for the communications technology landscape—so it worked for us. It allowed us to focus on more complex technologies than other views might have done. We turned away from early versions of instant messaging because we thought that they were technologically prosaic and mundane in terms of the richness of the (human) communication in question. We turned instead to the technologically more elaborate audiovisual systems to address what we saw as more profound issues of human communication. Our view led us to focus on complex bodily matters, where the bonds of communication are sealed by a sense of being together in real time and in shared space. We were less concerned with matters that might be measured in terms of human value or intention which transcend and are indifferent to time and space. The shared whiteboards were used for some fairly complex social actions (such as play and romance), but we viewed those behaviors as less interesting to invent for because they used only simple technologies and simple behaviors. In our judgment, they used only one mode of bodily action—the keyboard. Our other inventions seemed more appealing since they used multiple modes of bodily action as well as more elaborate technologies. We were drawn to multiple sensual dimensions in communication. These sensual dimensions were related to the senses of the body only, not to the senses of the mind; to what could be felt in the here and now. More dimensions of real time connectivity would demand more innovative technological solutions, we thought.

SQUEEZING HUMANNESS INTO BODIES

How does one judge such motivations—by comparing them against richer visions of the human or by simply counting how many inventions derive from using that view? These are not the same measure at all. We ought to be kind about these motivations and generous too in our measurement of them. They drove us to invent and produce a lot of IP—both invention proposals (the first documented stage in applying for a patent) and intellectual property. Nevertheless, paradoxes were observable in the circumstances of the work in question just as they were in the scope of the inventive drive.

As I noted, when the video tunnels experiments were completed, we packed up the enhanced system and went back to the old audiovisual system. There were some practical reasons for doing so. The new set-up was bulky, for example. But we also didn't imagine that we would find it useful for ourselves in our own activities. Its richer set of bodily fittings did not actually equal an easier mode of expression. Indeed, the contrary was true. Whenever one used the video tunnels or the simpler audiovisual system, bodily fittings—mutual glancings and lookings—were only part of the implications of the actions in question. When I shared a video connection with other people, they were interested more in why I wanted to glance at particular objects than in the objects themselves. Making a video connection was a powerful act. It was a kind of intimacy—a closeness that not all of my colleagues thought was appropriate. By *appropriate* here I am alluding to the moral implications of such acts, not their physiological ones. One motivation implied was that we were seeing if our colleagues were really working. This could be a management act, a kind of surveillance. Many colleagues resented such intrusions. Another could be to deepen a relationship. Indeed, one couple in EuroPARC became well-known for using the media space for precisely this activity. But because this was so, other attempts to use the media space became suffused with questions about whether such use was similarly motivated—an attempt to push the boundaries of work collegiality. But such concerns were beyond the scope of our inventive landscape, and we did not worry about them. When we did bother to look at them (over coffee or when we reflected at the end of the day), we saw that these concerns would be tricky to unpack and use as a basis for design. We recognized that issues of human action were relevant here, but our instincts were to avoid them; Turing's aversion to moral overtones encouraged us away.

If we had problems fitting the mind's landscape to the audiovisual system of EuroPARC, today we have even more problems fitting the mind to the latest systems that we have devised to allow interaction over distance. With the media space, we played with the infrastructure, but with our new interaction-rich conference systems, we don't even play. After experiments are completed, they are packed up and put away. My colleagues have no desire to use these systems for themselves. Again, they use the phrase *ordinary people* to suggest that others, somewhere else, might be willing to do so. But they view as uninteresting the question of why they themselves don't want to use them. Answers to that question don't lead them to think of any new inventions. It doesn't seem to have any pragmatic value.

This attitude reflects an indifference to intention and expression—a casual looking away from what one might label *mindfulness* in human action. To say again, I think Turing is a primary source of this. He deliberately chose to focus on the bodily side of human action so as to avoid considering the other side, the mind—or as I am suggesting, the expressive and intentional—and so too do many of my colleagues and I when we invent. Other things might be done with the technologies we invent, and human endeavors encompass more than we design for. But that more, whatever it might entail, didn't—and most often doesn't still—interest us when we get on with our work, our inventing business. We focus instead on what we think is a better route to inventiveness. One consequence of this is that it leads us to disregard technologies that might be more influential in the world at large (such as instant messaging). The audiovisual technologies that we invent seem to be introduced to a world that seems largely unimpressed. But we are impressed by our inventions. We are inventing for a richer human experience, and we think that this will keep us at the cutting edge of invention even if the people out there in the "real world" haven't yet come to adopt those technologies. We are inventing for the future of communication, even if that future never seems to end up turning into the reality our inventive spirits expect.

PLACING PRAGMATISM

Turing was not alone in developing the basics of this pragmatic view or in constructing visions of the human that divide mind and body. But both our use of his view and also Turing's motivation should not be treated as being philosophical or ontological (that is, as having to do with true measures of reality or in this case humanness). As I say, our concerns, like Turing's, were with other matters. For us, it was with invention. For him, it was to craft an understanding of what computer algorithms could do.[9]

Other thinkers in the 1930s weren't pragmatic in this sense and sought instead to offer visions of the human that they claimed were truer, more scientific. In psychology, for example, B. F. Skinner and various others developed what came to be called *behaviorism* at pretty much the same time. Behaviorism was fiercely opposed by many at that time and ever since. This period created a watershed between different sciences of the human. Sociology and anthropology started to oppose any attempt to split mind and body (their practitioners were particularly averse to behaviorism), for example, and have continued in their aversion. The leitmotif of one of the major sociological books of the early postwar period, Talcott Parsons's *The Social System* (1951), was critical of behaviorism, and though his alternate vision of the human is no longer fashionable in sociology, his opposition of behaviorism certainly is. These disciplines have also shown persistent aversion to what might be said to be the opposite side of behaviourism, a focus solely on the mind and its purported structures, most often assumed to be computational (such as structuralism associated with Lévi-Strauss in the 1960s and 1970s and the mentalism associated with Jerry Fodor in the 1970s and 1980s. See Lévi-Strauss 1963; Fodor 1975). Meanwhile, philosophy developed a response to Skinner and to Turing, especially as Turing in his later papers started moving from pragmaticism toward a kind of ontologizing. A colleague of Turing's at Cambridge University, Ludwig Wittgenstein, developed his philosophy in response to the kinds of reasoning that were underscored by Turing, creating a wholly different path of reasoning about what it means to understand human action. In his view, intelligence is not something that can be properly understood by an empirical test. It's a conceptual framework leading to ideas about culture, practice, rule following, and, yes, occasionally demonstrable proofs or tests. This is essentially what Wittgenstein's *Philosophical Investigations* (1952) is about. Turing, Skinner, and various other proponents of dualism may have persuaded some but not by any means all.

THE LIMITED HUMAN

If a concern with the body doing the communication has been central to the inventive landscape of the past twenty years or so, one of the correlates of this view is that the body is also a processing machine, one with limits. For many contemporary researchers on communication technology, that the human machine has limits and thresholds is the starting point of their endeavors. While my colleagues and I have sought to offer richer ways of supporting body interactions between humans, for interactions between humans and computers, and eventually, for interactions between humans via computers, others have in mind the idea that the human—the user—has specifiable limits of attention and processing, beyond which they cease to be efficient.

This view helps guide another set of research activities that I see around me. As I mentioned at the outset of the book, some of my colleagues seek ways of filtering communications so that only the really important are allowed to get to the top and to interrupt. Their technique involves gathering empirical data about human doings that are combined to produce rule-based systems that determine the kinds of messages that ought to be allowed through and the kinds that should not be allowed.

At first glance, these applications seem appealing. One can almost hear ordinary users shout out with glee at the prospect of having a tool that automatically sorts their emails and leaves them to deal with only the really important ones.

In these research activities, however, the model of the human that underscores how they do this business inverts the Turing theoretic approach. It concerns not the body but the mind—though there is an essential commonality here. This metaphysics is still dualistic and renders humanness quite peculiarly because of it. Both this view and the body-emphasizing view have been adopted largely for pragmatic reasons. Just as some researchers have looked at movements of the body to inspire their inventive imagination, this set of researchers has looked at what they think of as the mind (or the *brain*, a word they use interchangeably with *mind*) for theirs. The researchers in question have a somewhat elaborate notion of what the mind might be. It is a vision that includes certain notions about how the mind works that one might not assume from, let us say, a commonsense perspective. These notions are used to inspire a particular line of inventive inquiry. For example, pretty much at the same time as the media space was being developed at EuroPARC, a number of researchers developed a wholly different yet interdependent set of technologies that could allow computers to replicate what they thought was human memory. These researchers had the idea that the human mind wasn't merely like a computer but was rather a somewhat bad computer that suffered from retrieval problems. In this view, the mind had data in its stores but didn't have the right mechanisms to identify which data ought to be retrieved at any moment in time. Memory (or more exactly the problem of recollection) offered an example of this, and researchers assumed that the difficulties people had in remembering were proof of their computational inadequacy.

With this idea as a motivator, researchers built a system that created a visual record of what people did during a day (or at least what they did in the lab where the infrastructure worked), and these records could be replayed as prompts when recollection was required. The system functioned by having every user wear a badge that communicated to the embedded systems whenever a user sat down or moved anywhere within the building. This immediately prompted the audiovisual systems (the media space) to create a trace (a set of video segments) of that behavior. At the end of the day or week, the user in question could get the system to replay the video (see Newman et al. 1991).

Watching the resulting videos turned out to be fun. The viewing angles of the cameras used to collect images were often odd: heads would be missing from some views, and only the tops of heads would be visible from others. One could vary the speed, too, so that the video would play like a 1920s movie, with a kind of staccato performance of the actors in question (oneself and one's office colleagues). When set fast, the video made one look frenetic, when played slowly it made one look indolent. Despite this amusing side, some researchers claimed that it invaded their privacy, even though it was shown to be no more invasive than the media space without the addition of badges. What seemed more salient to their complaints was the fact that the badge system seemed to be an icon for what group or gang the researchers in question wanted to be seen as affiliated to—the badges outfit or some other.[10] The political issues notwithstanding, the system itself did not get used for the ordinary work of the lab or its staff. My colleagues did not feel that they had any need for the technology. As with the media space, this technology seemed designed for others, somewhere else, with other problems.

Some years later, a similar technology has been built in my current lab (see Hodges et al. 2006). In the former case, the data were captured by systems in the built environment (the media space and its associated networks and database systems), but this new approach captures the data in the devices worn by users.

It reflects developments that have been made in hardware in the time since the EuroPARC was built (the late 1980s). But the new technology has a similar set of ideas behind it, with some minor differences and refinements. Before, the mind was viewed as some kind of multi-element computer system with a database for memory and another system for retrieving memory data. Now, the mind is still viewed as some kind of container, but the view is augmented with the idea that the mind has internal displays (like a cinema screen)—qualia—of its external inputs. The technology offers a kind of external replica of these qualia—visual records of what the user or the wearer would see.

As with the prior research, a concern has been to see if the system can help address problems of memory, but here the question has been to offer a supplement to the dataset that is constitutive of memory rather than a mechanism for its retrieval. In this case, the devices in question (which are remarkably similar to the badges deployed in EuroPARC) have been designed to be worn by the user so that what they do can be captured, and the devices themselves do the capturing by automatically taking hundreds of images over the course of a day. The devices can take several thousand images before they are full and might last more than a day or two. Having caught the pictures, the technology has been designed so that users can download the images on to a PC and, with the right program, replay them or view them. As with the EuroPARC system, these pictures can be played quickly or slowly and make the activities of the subjects look peculiar in various ways. The staccato effect is still there, as are the orthogonal views (a head seen from below, a desk seen from an unusual angle, the device sometimes hanging down and almost touching the desk in front of a wearer). Some of my colleagues have used the devices to see what they are like, but none have ended up using them as part of their routine activities. As with prior systems, the assumption that seems to have motivated this research was that the devices would help others—others who might have memory problems derived from senility or another form of neurological decay—but not the researchers who devised the technology.

One or two researchers (myself included) have investigated whether the devices might be used as wearable cameras and not as things intended to aid the brain.[11] Our research into gathering pictures in this manner (by having a wearable device that automatically captures images) has found that people use such pictures not as mirrors of the past (as qualia) but as vehicles to exercise their imaginations. They use them to look anew at what they recollect, to see it in a different light, or even to see and discover aspects of their affairs that they would normally neglect.

The lesson we took from our research was that, in normal life, a person's sense of the past and of memory is not thought of (by them, our users) as a trail of material stuff that is collected by their minds. In their view, the past is a place that they looked at with fresh eyes and often differently when they recalled it, depending on their purposes for recollecting. Our study participants used the devices to help recall the past, but their purposes (to tell a story, perhaps) did not encourage the idea of memory as a container that is gradually filled up. None of these concerns would lead one to imagine that the head has a series of films (qualia) inside it replaying. We came to see that recollecting moments from the past—moments selected from further away or nearer in time—was better thought of as a constructive process than as a computational one. The past seemed to be a much more complex place to visit and comprehend than the vessel metaphor (and the qualia concept) would allow.

Some philosophers have difficulty accepting the idea of qualia. As Norman Malcolm notes in *Memory and Mind* (1977), the qualia thesis cannot be accepted as true for the simple reason that there must be a qualia for every event, including memory events. Each time there is a recollection, another qualia will be created, and each time someone recalls that recollection, there will be another in turn, ad infinitum. Ultimately, the qualia concept suffers from the problem of infinite regression.

But my colleagues don't worry themselves with these philosophical doubts. Because they are constructing a view of the mind that is entirely pragmatic, not philosophical, they disregard the problem of recursion; for them the idea of qualia is a pointer toward inventions that might solve something. They have come up with the idea that they might control qualia through filtering. In this way, one would not keep producing endless qualia that would fill up the mind (so to speak), but one would present qualia that summarized and triaged the past. They have been seeking to devise techniques that will recall only what matters. To do so, these researchers haven't confined themselves to visual traces of action (as the original badge-based system did) and have sought instead to try to assemble as much as possible about the inputs that humans manage.

For example, in an early but nevertheless good introductory paper, "Models of Attention in Computing and Communication: From Principles to Applications," Eric Horvitz and colleagues (2003) explain how to use sensors to create the stuff that constitutes qualia and then how to sort or triage this stuff to provide only what counts.[12] The sensors in question include "microphones listening for ambient acoustical information or utterances, cameras supporting visual analysis of users' gaze or pose, accelerometers that detect patterns of motion devices and location sensing via GPS and analysis of wireless signals . . . , online calendars and considerations of the day of the week the time of the day" (54). This material is then analyzed through probalistic attentional models that the researchers hope can determine what the users really need to recall. Basically, this involves measuring the frequency of incidents at some moment in time and the frequency of the recurrence of those same incidents through time. These statistics produces weightings that allow the system to distinguish the relative importance of incidents.

This is well away from the problem of glancing and touching, of looking and peering, that we saw with the media space system. It might seem well away from communications, too, but when one can see easily how this view can lead researchers to address the problem of communications overload. In this view, the a user manages multiple mental tasks and is concerned about reducing the burden that these tasks impose: people think of themselves as machines, in other words, and worry about optimizing their performance. Technologies derived from this point of view help them in this.

For example, some early attempts to do this looked at home life. Here, the models were designed to ascertain what would be a good time for a message to be received, to interrupt whatever the person was doing, and when would not be. But these models didn't work, since people's preferences were not only idiosyncratic, with unique rules and requirements, but also changeable—a person's mood seemed as important a factor determining whether someone wanted to deal with a message as any other more objective measure (like the intensity of the other activities they were engaged in).[13]

More recent research has focused on work settings, where the problem of determining whether a communication is an interruption is made by comparing the nature of that communication to the tasks that the recipient of messages is currently engaged in. These tasks are captured by accessing the person's activities on their PC. If they are preparing a spreadsheet then a message will only be allowed through if it pertains to the topic of that spreadsheet (assuming that the title gives some indication of topic); if they are Web browsing in relation to some project then the "interruption management system" will only allow messages about that project through, and so on.

Unfortunately, although these systems sound appealing, in practice they rarely go beyond prototype. Even the inventors of the systems admit that they don't like using them much themselves. They miss the interruption of messages, they explain; when they use the systems, they feel as if they are getting "out of touch." Besides, outside of lab settings, it is almost impossible to gather all the data required for the systems to work. It is little wonder, therefore, they never see the public eye.

Whereas these systems seek to inhibit messaging, other system simply triage, allowing the user to choose how many they want to attend to at any time, but offering hints as to which is most important or urgent. These systems also have problems. They tend not to succeed with person to person type messages, since it turns out there is no effective way of identifying importance. For example, some systems use the identity and status of the sender as a criteria to distinguish between importance; others the frequency of messaging; some do both.

But these criteria don't work in a way that pleases the users. With these systems, mail from the boss gets through but infrequent email from a colleague does not; messages from those who message too much do. The subtle patterns of social action that give messages their meaning gets lost. Again, these kinds of systems never get beyond prototype.

As a result of these failures, some researchers have turned to blog postings with a view to helping filter these in a way that lessens the burden placed on the person trying to keep up with them. At the current time especially, there is an interest in analyzing Twitter feeds and selecting from them only those that the users "really want." The systems do this by storing and analyzing prior selections the users have made, and "learning" from these "click-throughs" want the user will prefer. These systems also look like failing. It is not that sometimes those who use Twitter want to read what has been said on a particular subject. It is rather that part of the charm of the blogosphere is accessing it leads to the discovery of new topics and threads. As we saw in chapter 2, it is precisely because these topics change that people turn to the blogosphere to keep up to date. Systems that use predetermined topics to select content miss the point. Though they reduce the amount of content sent to the user, they suffocate the desire that gives life to the act of blogging in the first place.

THE METAPHYSICS OF OVERLOAD

The perspective that has motivated these research endeavors brings to mind many recent books that aren't about acts of communication but are more concerned with our information saturated lives, such as Richard Lanham's *The Economics of Attention: Style and Substance in the Age of Information* (2006). This is an exposition by a literary theorist of how an economic perspective might be applied to the problem of searching, navigating, and finding delight on the Web. But arguments about humans as information processors—as machinelike entities with limits—are old hat. For example, Norbert Wiener's 1948 book, *Cybernetics: or the Control and Communication in the Animal and the Machine* is a manifesto of a view that emphasizes the idea that a human body is an information-processing machine and in particular a processor of communicated signals. This sounds very much like the modern psychological theory that Horvitz, Nagel, and others evoke; it also sounds very pertinent to our current concern.

Wiener argues that the stuff people process (like the words conveyed in a message) is like any other kind of stuff that the body might receive and produce; as material to be processed. He argues that the processing machine (the human) has certain key properties. In his view, processing of stuff, whatever that stuff is, must stabilize or else the system (the human) will break. Wiener proposed that this stabilization—this performance, as he put it—could be measured quantitatively. He offered in *Cybernetics* (and elsewhere) mathematical techniques and concepts that he thought would help to measure this information processing so that people could predict processing optima and processing stability.

In the 1940s, many people found this view profoundly appealing. They came to think of cybernetics not as a label for a point of view but as a science of predictive models of human behavior. The echoes of this particular claim were heard across many disciplines and were so loud that universities set up departments of cybernetics to investigate what the human processing machine might be. Many disciplines viewed this as a key turning point in their development. Ergonomics, for example, saw cybernetics as solving how a symbiosis of man and machine might be analyzed. With Wiener's ideas at its heart, it came to describe human-and-machine systems in which the two could work together harmoniously by optimizing the processing burdens of each. One set of stuff to process was what a machine was good for, and another set of stuff to process was what the person was good for.[14]

This sounds very much like the approach taken in a contemporary discipline, Human Computer Interaction, or HCI.[15] Cybernetics and modern HCI do have something profoundly in common. Wiener's view seems to be essentially the view of many of my contemporary colleagues (whether or not they would admit it), particularly those addressing issues of overload. But it is also closely related to the views of those who take a more Turing theoretic position. Indeed, the view of the human in both perspectives merges in a consequential way in the landscapes I am describing — the inventive world of corporate research. In Wiener's view, the user is treated as a machine of sorts—as a processing machine. But the user is also thought to be something that exists only in real time—what it processes is only inputs and outputs in the here and now. And it is this concern with space and time that is common with those who emphasize the body.

In terms of the technologies that this combined vision produces—a lot of it is very innovative and useful. It has led to novel ways of communicating across distance and healing the apparent misfit of perspectives that early communication systems generated; it has lead to richer ways of interacting across distance. The vision has also lead to interesting approaches to the problem of overload, offering various techniques that calculate when too much is too much. All of this points towards new ways of communicating that heal the problems of distance and separation and do so without demanding too much of the user.

Nevertheless, this combined vision has looked at acts of communication with a reductive lens. Those who invent under the auspices of a Turing theoretic point of view or a Weiner like information processing one, only concern themselves with some kinds of communications acts, not all. Indeed, if we learn anything about the value of letters and the sensibility for being in touch that they cultivated discussed in chapter 2, then this vision is emphatically missing important concerns. This perspective lacks interest in material that is somehow beyond processing in particular moments in time or space. There is very little mindfulness in the human machine that communicates here. For example, aspirations and hopes for the future don't matter for the human in the centre of this vision; these are not substances that can be processed in the same way that real time input and output signals can. They cannot be fixed in some spatio-temporal location (though where someone is will often help cultivate them). Similarly, recollections and laments about the past do not fit into this vision either. Nor do the feelings that one cultivates for another through acts of communication, especially if those feelings are beyond or distinct from those created through activities that entail "being together in the here and now." Digitally created analogues for togetherness that the body centric and information processing view produce, namely those created by Wave, by media space technologies and so on, don't allow for these elements of the communication acts to emerge. This is hardly surprising; they were not designed with them in mind.

In sum, the Turing theoretic and Wiener cybernetic vision does not allow inventiveness to address some of the metaphysics of togetherness we mentioned in chapter 2. As we saw with blogs, the desire to share in dialogues about the hubbub of the moment, to feel albeit for a fleeting moment part of the digital crowd, is not encompassed, for example. The efforts of my colleagues and I have not changed the landscape of communication in ways we expected or hope. Although some of our ideas do manifest themselves in technologies that get widespread take-up, most don't.

Besides this disappointment, there is a discontinuity between our thinking at work and our thinking elsewhere, when we abandon our work and professional hats. What is thought of as the limits of expression, as the limits of our efficiency from this Turing-Wiener view, does not equate with the everyday human measures of expression and expressiveness that we apply when we go home. Nor does our work vision allow us to comprehend and explain the criteria that we ordinarily use when selecting between, say, the written word or a videocall at home. In my professional world—in Wiener's and Turing's world—users would choose video conferencing if they were able to and if it doesn't overload them. Efficiency in this view is all about getting as much as possible, given particular cognitive limits. But at home, few of my colleagues or I often use Skype or other video-mediated communications. We do use them, but not always and indeed not often. But our work selves cannot account for this. Choosing to make a video contact in the domestic sphere is not made on the basis of how this mode of communication provides a richer array of sensual fitting—seeing as well as hearing, gesturing as well as speaking. It is chosen in large part because it makes the act of communicating special in itself. One doesn't use a video to call merely to communicate, but to make the act of communicating special—and, in so doing, making the parties involved special. Skyping turns out to be one of the ways that distributed families constitute a sense of being a family. It's not what is said on a video call that matters, it's the mere doing of it that does (see Ames et al.'s 2010 paper, "Making Love in the Network Closet: The Benefits and Work of Family Videochat"). That we do not appreciate these subtleties when we have our work hats on reflects the limited way we use the pertinent terms. At work, the expression *amount of words* is simply a synonym for *volume*, not a measure of adroitness, thoughtfulness, or neglect.

Turing believed he was inventing a new discipline, one that dealt with algorithms. But this vision also included a view of the human. As it happens, Wiener thought that the science he was inventing, cybernetics, was all about people, even though his science was enormously mathematical, and hence quite close to what Turing thought he was doing. But the world view that these individuals have produced is one in which people—the users—turn out to be not very human at all. They have human like capacities and human like behaviors to be sure, but they are so reduced in their sensibilities that the humanness has been taken out. The performers of communication acts are like robots or animals. We can see that this is so because when people complain about too much email or too many postings on their social networking sites it is precisely because of things bound to the human condition, to the sensibility that the metaphysics of communication has produced for us; animals and machines can neither sense that or understand it. It is not about space, time or information processing. It's something more, something greater; about being in touch but when being in touch is a moral matter, not a physical one.

CONCLUSION

Based on the arguments presented in this chapter, it should be clear that the world that I live in, like any organizational world, cannot be easily mapped out like the geography of a country. But it is a landscape of sorts. It has certain salients that, once described, can provide a sense of what it might be like to move around within it. Some places are commonly investigated, and others less so. Some domains seem almost beyond the pale. This landscape has a kind of unity or general patterning that gives it particular form. But one might also say that this landscape is not systematically laid out by any plan or map. Although this landscape may be the place my colleagues and I traverse and although its shape might lead us in some directions and not others, our sense of this place as we go about our daily lives is not perfect or clear. My intellectual landscape is like all landscapes. I navigate through it with routine and habit, and then occasionally my sense of it is disturbed by moments of reflection and doubt. There is a sense sometimes that one is seeing something new—a vista beyond what has seen before or a wood that is at last recognized, since we have somehow escaped from the trees.

Despite sometimes getting lost, my colleagues and I have a notion that what we are doing and where we are going does have some sense or purpose; a direction even. We are afflicted by a conviction that somehow and for someone (our colleagues and our managers perhaps), what we are about does indeed have reason. This landscape and our personal convictions about our trajectories through it have led us to invent not just anything. It has led us to explore technologies of a particular kind, ones that reflect what we have come to think of as the geographies of human communication. In our landscape, the human is a kind of body, a machinelike body, and the nature of its communication is machinelike, too, with thresholds, limits, and processing requirements, all of which are fixed in particular spaces and times. Communication is about bringing machinelike bodies together, across or through space, without overloading their systems.

All this might seem a long way from the questions asked at the start of this chapter and this book: why do we keep inventing so many new communications technologies, and why do we complain about them even as we invent them? But there is a link, and the link is a paradox that has to do with how the professional world I have described—my own world—leads members of it toward an end point. Our very practices of inventing for that end point create demonstrable proofs that the human model we have is not the one we orient to in our worlds outside the labs. What we invent, we don't use. Though people on the outside might take up elements of what we invent, we don't get excited by those applications, thinking them too feeble to be worth investing in. We invent for one world and live in another.[16]

The examples provided in this chapter indicate the direction of the inventive imagination behind research into new communications. The direction leads us along a path where more is viewed as better. The fate of the shared whiteboards illustrates this. If they allowed a number of persons, separated by distance to see what each other wrote, then we imagined that offering them the ability to see their correspondents would be better, more appealing, a closer fit to the geography of their natural communication acts. Hence we turned to the media space and neglected the whiteboards. And if one has the written word and sight, why not also have gesture? Hence we turned to C-slate. Our moves were intended to take the user from a monosensual mode of expression to a multisensual one; from impoverished geographies of interaction to richer ones.

The examples show how this view, a credo, if you like, leads my colleagues and me to invent applications, devices and technologies that are designed to allow the communicating, processing body to do more. I have noted that we have been fumbling, but persistent, in our efforts to do this. Our understanding of what this more might be is bound up with our vision of the human—which emphasizes action rather than intention and quantification rather than quality. This credo also has a notion of limit, too, which is our goal. We orient our designs to an endpoint when the user will have enough—at time when a system or set of systems will offer all that is needed or all that a person can handle. We are, after all, machines, and like all machines, we have limits.

So why do we keep inventing new communications technologies as we complain about juggling too many communications? And yet even more muddling, why do we seem to ignore our own complaints and use some of the old technologies that our inventions are intended to replace? It's not simply that the new technologies aren't available; something about the old ones appeal. Our work selves hold a view of the communicating human communication and our private selves have another.

It seems to me these private views are richer, more subtle, more accurate. This private self understands that communication is not best thought of in terms of volumes, capacities, and scopes—the view that our work selves deploy. Our private selves are charmed by the different experiences that communications channels afford, not by the way they offer more sensualities (sight, sound, and touch, for example) but by how they broaden our expressive repertoire. Our private selves know that this repertoire is not to be indexed by behavioral geographies alone. In our private lives, we are deeply familiar with the fact that there are many dimensions to expression—variety, depth, lightness, spontaneity, and ease. We know, too, that we can find enjoyment in some channels because they are private and find enjoyment in other channels because they are public. And we know as well that a withdrawal from communication can allow us to recast our intentions in another channel. We know that a letter, written alone and diligently crafted, can say much more than a videocall could ever allow us to say. In sum, our private selves recognize that *more* in this sense has a meaning that is quite unlike what the word *more* means when it is used in our inventive endeavors at work.

Here lies one of the paradoxes of our contemporary communications age—how the credo used in certain inventive landscapes that have helped produce some of the communications infrastructures we see around us is so impoverished when compared to the views of the human captured in, evoked by, and oriented to in ordinary reason and everyday language. I am thinking here of everyday folks getting on with their ordinary yet complex lives and using communication means as opportunities to manage their diverse affairs. Our scientific selves seem to invent for a future that is expected to be populated with humans that are somehow much less than this, and though these people are more machinelike, this is not because they are complex but because they are simpler and as a consequence less human because of it.

And if that is so there is a greater question that follows on from this. If the technologies that this credo leads to don't get used by the world at large as expected, how does one account for the technological landscape? that does exist? If the inventors can't be said to produce it, who has? Is an interplay between the products of invention and the desire of the users that shapes the technological landscape? The question that motivated this chapter was who is the user designed for? It may be that we have been looking in the wrong place. We might want to look at practices rather than at invention, at what people do rather than at what those in the inventing trade think people do. It is to that possibility we now turn.

NOTES

1. The mobile phone and network manufacturers have research labs in this space. Mobile phones have become small, handheld computers, not a different technology species from PCs as they used to be. Similarly, the networks that support them are essentially vast, air-based Ethernets. Whatever their organizational provenance, such places are not the only drivers of change of course. Be that as it may I talk about Google research later on in this chapter.

2. The approach I take is not the only one that can be used to explore the media space either at EuroPARC or in other establishments where versions of the technology were being deployed. The best introduction to the many research endeavors undertaken is to be found in Steve Harrison's 2009 book *Media Space: Twenty+ Years of Mediated Life.*

3. See my own studies on this topic: "Looking at Ourselves" (1992) and on the role of technologies to symbolize identity, "Why People Do or Don't Wear Active Badges" (1996).

4. This question is often asked from a sociological perspective. For a good introduction to this point of view, see Keith Grint and Steve Woolgar's 1997 book, *The Machine at Work.*

5. For more information on this and other developments of the media space see the second chapter in Harrison's book, *Media Space,* 2009.

6. The system I have in mind has not lead to any publications on this topic, although it is often bundled under the category of projects called i2i on the Microsoft Research portal.

7. See http://wave.google.com/help/wave/about.html.

8. One could obviously spend a great deal of time assessing whether this is really so—the biographical approach I use is not the only way one might judge on this. For a similar approach but with a different technology in mind, see Agre's 1997 paper, "Towards a critical technical practice: Lessons learned in trying to reform AI."

9. Turing's view of the human in the 1936 paper also allowed him to do *his* work—to define some parameters of computer algorithms. That was the focus of his thinking, not an ontology of the human. In later years, by his 1950 paper, for example, Turing did drift in to that concern but that is another matter.

10. See my 1996 article on badges.

11. See, for example, Harper et al. 2007 and Harper et al. 2008.

12. This paper is simply representative of the perspective in question; there are many papers on this topic, including those by the same authors. Another good introduction can be found in "BusyBody: Creating and Fielding Personalized Models of the Cost of Interruption," by Eric Horvitz, Paul Koch, and Johnson Apacible (2004).

13. An early academic paper reporting on such efforts is Nagel et al.'s 2004 study, "Predictors of Availability in Home Life."

14. Not that this is easy, even with this model. See Bainbridge's paper, "The Ironies of Automation" (1983).

15. The canonical text here is Card, Newell, and Moran 1984.

16. Although I am presenting a biographical argument, my own experiences represent what gets done in many research labs. For corroboration of the claims I make about the emphasis on the body, see, for example, Kjeld Schmidt's dispirited review of research titled "Divided by a Common Acronym: On the Fragmentation of CSCW" (2009). He finds that throughout many of the communities investigating collaborative work (such as CSCW, computer-supported cooperative work) there is tendency to reduce the complex forms of human collaboration into the simplicities of temporally fixed body mechanics. See also De Vries 2005.

REFERENCES

Agre, P. E. 1997. Towards a critical technical practice: Lessons learned in trying to reform AI. In G. Bowker, S. L. Star, and W. Turner, eds., *Social Science, Technical Systems and Cooperative Work* (131–157). Mahwah, NJ: Erlbaum.

Ames, M., J. Go, J. Kaye, and M. Spasojevic. 2010. Making love in the network closet: The benefits and work of family videochat. In *Proceedings of CSCW2010, Savannah.* Chicago: ACM Press.

Bainbridge, L. 1983. The ironies of automation. *Automatica* 19(6): 775–779.

Bittner, E. 1977. "Must we say what we mean?" Communication and social interaction. In P. F. Ostwald, ed., *Communication and Social Interaction* (180–197). New York: Grube and Stratton.

Card, S., A. Newell, and T. Moran. 1984. *A Psychology of Human Computer Interaction.* New York: Erlbaum.

De Vries, I. 2005. Mobile telephony: Realising the dream of ideal telephony. In L. Hamill and A. Lasen, eds., *Mobile World: Past, Present and Future* (11–29). Goldalming: Springer.

Fodor, J. 1975. *The Language of Thought.* Cambridge, MA: Harvard University Press.

Gaver, W., T. Moran, A. Maclean, L. Lovstrand, P. Dourish, K. Carter, and W. Buxton. 1992. Realizing a video environment: EuroPARC's RAVE system. In *Proceedings of CHI 92* (335–341). Amsterdam: ACM Press.

Grint, K., and S. Woolgar. 1997. *The Machine at Work: Technology, Work, and Organization*. Cambridge: Polity Press.

Hacker, P. M. S. 2007. *Human Nature: The Categorical Framework*. Oxford: Blackwell.

Harper, R. 1992. Looking at ourselves: An examination of the social organization of two research laboratories. In *Proceedings of CSCW '92, 31st Oct.–4 Nov. (Toronto)* (330–337). New York: ACM.

Harper, R. 1996. Why people do and don't wear active badges: A case study. *CSCW: International Journal* 4: 297–318.

Harper, R. 2005a. From teenage life to Victorian morals and back: Technological change and teenage life. In P. Glotz, S. Bertschi, and C. Locke, eds., *Thumb Culture: The Meaning of Mobile Phones for Society* (101–113). Bielefield, Germany: Transcript Verlag.

Harper, R. 2005b. The moral order of text: Explorations in the social performance of SMS. In J. Höflich and J. Gebhart, eds., *Mobile Communication: Perspectives and Current Research Fields* (199–222). Berlin: Peter Lang GmbH–Europäischer Verlag der Wissenschaften.

Harper, R., and L. Hamill. 2005. Kids will be kids: The role of mobiles in teenage life. In L. Hamill and A. Lasen, eds., *Mobile World: Past, Present and Future* (61–73). London: Springer-Verlag.

Harper, R., D. Randall, N. Smyth, C. Evans, L. Heledd, and R. Moore. 2007. Thanks for the memory. In *Interact: HCI 2007*. Lancaster: British Computer Society.

Harper, R., D. Randall, N. Smyth, C. Evans, L. Heledd, and R. Moore. 2008. The past is a different place: They do things differently there. In *Proceedings of DIS (Designing Interactive Systems)* (271–280). New York: ACM Press.

Harrison, S. 2009. *Media Space: Twenty+ Years of Mediated Life.* Godalming: Springer.

Hodges, S., L. Williams, E. Berry, S. Izadi, J. Srinivasan, A. Butler, G. Smyth, N. Kapur, and K. Wood. 2006. SenseCam: A retrospective memory aid. In P. Dourish and A. Friday, eds., *Proceedings of Ubicomp 2006* (177–193). London: Springer.

Horvitz, E., C. Kadie, T. Paek, and D. Hovel. 2003. Models of attention in computing and communication: From principles to applications. *Communications of the ACM* 46 (3):52–59.

Horvitz, E., P. Koch, and J. Apacible. 2004. BusyBody: Creating and fielding personalized models of the cost of interruption. In *Proceedings of CSCW04* (507–510). Chicago: ACM Press.

Huurdeman, A. A. 2003. *The Worldwide History of Telecommunications.* Hoboken, NJ: Wiley.

Izadi, S., A. Agarwal, A. Criminisi, J. Winn, A. Blake, and A. Fitzgibbon. 2007. C-Slate: Exploring remote collaboration on horizontal multi-touch surfaces. In *Proceedings IEEE Tabletop*. Newport, RI: IEEE.

Lanham, R. 2006. *The Economics of Attention: Style and Substance in the Age of Information.* Chicago: University of Chicago Press.

Lévi-Strauss, C. 1963. *Structural Anthropology*. New York: Basic Books.

Maffesoli, M. 1996. *The Time of the Tribes: The Decline of Individualism in Mass Society*. London: Sage.

Marr, D. 1983. *Vision: A Computational Investigation into the Human Representation and Processing of Visual Information.* New York: Freeman.

Malcolm, N. 1977. *Memory and Mind.* Ithaca: Cornell University Press.

Nagel, K., J. Hudson, and G. Abowd. 2004. Predictors of availability in home life context-mediated communication. In *Proceedings of CSCW04* (497–506). Chicago: ACM Press.

Newman, M., M. Eldridge, and M. Lamming. 1991. PEPYS: Generating autobiographies by automatic tracking. In L. Bannon,

M. Robinson, and K. Schmidt, eds., *ECSCW 1991* (175–188). London: Kluwer.

Noë, A. 2009. *Out of Our Heads: Why You Are Not Your Brain and Other Lessons from the Biology of Consciousness.* Cambridge, MA: MIT Press.

Parsons, T. 1951. *The Social System.* New York: Free Press.

Piccinini, G. 2003. Alan Turing and the mathematical objection. *Minds and Machines* 13:23–48.

Schmidt, K. 2009. Divided by a common acronym: On the fragmentation of CSCW. In I. Wagner et al., eds., *ECSCW 2009: Proceedings of the 11th European Conference on Computer-Supported Cooperative Work, 7–11 September 2009, Vienna, Austria, 2009* (223–242). London: Springer.

Shanker, S. 1998. *Wittgenstein's Remarks on the Foundations of AI.* Abingdon: Routledge.

Sheutz, M., ed. 2002. *Computationalism: New Directions.* Cambridge: MIT Press.

Sudnow, D. 1972. *Studies in Social Interaction.* New York: Free Press.

Turing, A. 1936. On computable numbers, with an application to the Entscheindungsproblem. *Proceedings of the London Mathematical Society* 42: 1936–1937.

Turing, A. 1950. Computing Machinery and Intelligence. *Mind* 59:433–460.

Wiener, N. 1948. *Cybernetics: Or the Control and Communication in the Animal and the Machine.* Cambridge: MIT Press.

Wittgenstein, L. 1953. *Philosophical Investigations.* Trans. A. N. Anscombe. Oxford: Blackwell.

4 PARADOXICAL DELIGHTS

PREAMBLE

A particular view of the human—that expressive actions are to be grasped by the interlocking of lookings and glancings and that people can communicate only a certain amount of information because of limits imposed by bodily and mental processing powers—reminds one of cyborgs. But these cyborgs are already in front of us in flesh and blood and biomechanics. They are the humans of Andy Clark's *Natural-Born Cyborgs: Minds, Technologies and the Future of Human* (2003). This analogy between humanness and cyborg humanness lies at the heart of this chapter. What has been argued thus far in the book is that this cyborg vision cannot be adequate. It's too reductive, emphasizing the body too much and intentionality too little and treating expressive capacity not as human artfulness but as volume. Although some might claim that treating humans in a reductionist fashion is how certain sciences operate, reductionism of this sort does not lead to insights into our questions. Indeed, it can lead to distortion. Although the inventive spirit can find this reductionist view helpful, human acts of communication cannot be reduced to something else, like data exchanges between machines.

Can other approaches offer a better or more useful view? Turing's vision might be too pale and Weiner's vision too mechanistic, but what about views in, say, the humanities or the social sciences, do these attend to the sorts of questions I'm asking? I don't survey them all here, but for a number of reasons—partly biographical (I was trained as a sociologist) and partly intellectual—I attempt to come to grips with one of these views, sociology.

At first glance, sociology seems to be a natural approach to take since it appears to address the topic of this book. *Organizing* and *making a society* sound like conceptual synonyms for *being in touch*. But closer examination might bring this in to doubt. *Social structure* doesn't sound like *human intimacy* (which Henkin describes), for example. Nor does the phrase *social structure* suggest that sociologists are able to explain the passion that seems to motivate bloggers. One doesn't think of class when one reads a blog and absorbs its earnestness. Nevertheless, I turn to sociology because I think it can provide some (not all) answers to my queries. It does so because its topics include some of our own, even though some of those concerns get obscured.

In simple terms, sociology is preoccupied with a basic change that has to do with *community*. In sociology, the change in question always unfolds in one particular way: community is moving from the small scale to the large, from the rural to the urban, from being socially cohesive to being individualized. Individuals spend less time in face-to-face situations with extended family and friends and more time interacting with strangers across distances. The extended family is slowly reducing to the nuclear, and even these weakened bonds of kinship are gradually being replaced by ties of economics. Ultimately, community is changing from being local to being global.

Numerous phrases are used to label this dynamic, including the shift from *organic* to *mechanical* (Durkheim), the shift from *gemeinschaft* to *gesselschaft* (Tönnies), and perhaps most famous, the shift from *feudal* to *capitalist* (Marx). Although each of these authors had a slightly different interpretation, all viewed the change as something to be regretted: people are becoming more isolated, and the fabric that holds people together in a community and in something called a society is unraveling. This lament was summed up nicely by the phrase *bowling alone*, which was also the title of a 2000 book by the American sociologist Robert Putnam.

More recently, claims have been made that new modes of transport, their supporting systems (air and road, especially), and new computer and telecommunications technology have provided further impetus to this dynamic, thus accelerating the impact and scale of change. According to many, the geography of community is now being altered in ways that were not possible in earlier eras—through enabling a community to spread itself out across distance and in so doing reducing time as a constraining factor on social action (the time spent traveling to meet others is now less, for example). According to this view, community is being transformed into the myriad forms that exist only in terms of the connections (or *information flows*) between remote individuals. The ties of time, distance, and geography that linked individuals and community have been loosened by car traffic systems, mass air travel, the Internet, and mobile phones. Now people can create and sustain communities any place and any time and in so doing change what communities (and their roles within them) might be.

For most sociologists, including Nigel Thrift and John Urry, society has become more mobile than ever (see Thrift's 1996

Spatial Formations and 2004 "Movement-Space: The Changing Domains of Thinking Resulting from the Development of New Kinds of Spatial Awareness" and Urry's 2007 *Mobilities*). For others, mobile technology has created the most radical possibilities of all. Howard Rheingold argues in *Smart Mobs: The Next Social Revolution* (2003), for example, that the potential for change is now so great that all of us need to think carefully about how we let this technology affect our communities and what we want to be within them. If new mobile technologies are allowing individuals to be stripped of the bonds of space, time, and community, he asks, then what are the communities of which we are a part to become, and what are *we* to become? New codes of social conduct will be necessary, and the landscape in which we live will alter. Urban spaces will be transformed by the breaking of the relationship between space and action, for example, and governance will need to be rethought too. With "smart mob" technology, Rheingold tells us, it will be possible for enormous amounts of information to be provided about individuals, and this can be used either for the benefit of those individuals or for their surveillance. This is an argument that has been taken up more recently by Manuel Castells in his book *Communication Power* (2009).

Sociologists clearly see drama and much of it has to do with communication, expression, and being in touch. In this chapter, I present some empirical evidence that will allow me to explore what some of the changes in communicative practice might be. Many of these changes have to do with what some often dismiss as the details of communicative practice. Although these might seem small, they are as significant to the people they affect as are the centuries-old shifts from social to individualized experience, from the extended family to the nuclear, from social solidarity to social fragmentation, from fixed geographies of sociality to dynamic spatial flows of human interaction. These changes have more to do with such things as the social rules of turn taking in conversation than, say, the loss of identity or social belonging and the loss of fixed geographic connection in a chaos of movement. They have to do with such things as how embarrassment may occur and may be avoided, how some people seek to be where the action is through keeping in touch (in the blogosphere, say), and where boys and girls learn the painful arts of going out and being "an item" long before any matrimonial commitment is cemented in bureaucratic registries. They have to do with how technologically mediated forms of communication allow the sharing and giving of daily gifts in new ways and how doing so lets people build lives together at work, at home, and at play.

Our contemporary age is not only rendering the performance of human expression into entirely new forms. It is also enabling some patterns of expression that have been in existence for quite some time to persist and others to evolve in significant ways. Sometimes these changes are delicate and sometimes more visible; some finesse ordinary, everyday processes of human communication and some redefine its properties.

I explore these issues by reviewing the arguments put forward by one of the main sociological commentators on the nature of community in our time, Barry Wellman. According to Wellman (and many sociologists before him), place, time, individual, and community are being decoupled, and society is dissolving and emerging into something different—and this is being done through new modalities of communication, amongst other things. I examine his claims about a range of technologies (the Internet, mobile phones, and personal computers) and review in detail the sociological literature on the social effects of one technology, the mobile phone. This approach provides sufficient evidence for me to show that some of the changes that Wellman concerns himself with might indeed be appearing. It also allows me to present research that is focusing on some of the details that are also changing, which Wellman (and others similarly focused) ignore.

Various papers from the mobile-phone literature confirm some of Wellman's views, commenting on how *absent others* are now affecting social action in the caller's context, for example, and how the performance of dealing with the absent other has various forms and manifestations around the world. Being connected to people elsewhere is now a cultural necessity that is deeply embedded in contemporary codes of fashion and public behavior; it is intrinsic to social relations. But dealing with the absent other is only part of a much more complex matrix of communication acts that are immersed in considerations of the local, the immediate and the here-and-now. Both these local and remote relations are enabled by mobile phone technology. But we will see that some communication is not to be thought of as either remote or local; it is bound to the ways people create a sense (for themselves and those around them) that they are more than simply bodies sending whispers to those near and bellows to those far away. Somehow they exist beyond (or outside of) time and space in ways that are uniquely human; certain sorts of communication act enable this.

These topics provide me with grounds for exploring some data that my colleagues and I have gathered in various projects over the years. These data help me shed light on who are being brought together with mobile technology and provide insights into what they do when they connect. These are not differences between macro changes (those on a societal level) and micro changes (individual changes) but changes reflecting how people use modern communications systems to help fabricate the social bonds constitutive of their lives. The world is new, and the bonds that people make (and which make communities in turn) are central to this newness. But the bonds in question are made in ways that have to do with the communicative texture of our human lives. New technologies are central to this, but so are much more profound elements of the human desire, fear of, and practical adroitness with, being in touch. My chapter will address why saying hello, for example, can have complex ramifications and why, therefore, people will avoid responding to an act of communication. I will explain also why mobile phone technology is sometimes used to keep people apart. As we shall see, some acts of communication are difficult to do when face to face. Mobile phones allow people to avoid these difficulties.

My overall aim will be manifold. First I will show that if one can discern these complexities in the social shaping of one technology, the mobile, then it is almost certain that similar complexities will be found in examinations of other communications media. We will see more clearly that the view on the human communicator used in the world of technological invention misses many aspects of being in touch. Nevertheless, if we are confident in thinking that it is the human user that does the shaping of the technology, we must be alert to the richness and diversity of what this entails.

Much of the evidence presented about mobile phones is dated. As we shall see, when mobiles first became widespread, sociologists predicted that they would undermine elements of social cohesion. These fears were misplaced. But I take from this not that sociology was wrong but that when new modes of communication show themselves such fears are common. Though sociologists might think they are analyzing *gemeinschaft* and *gesselschaft*, ordinary users of new communication technologies also worry about the impact and shaping new communication technologies will impose on them. If the past decade commenced with fears about mobile telephony, the start of the second is concerned with the divisive impact of social networking, for example.

FROM DOOR TO DOOR TO ROLE TO ROLE

In an ongoing series of papers and edited collections, Barry Wellman, based at the University of Toronto, has explored the changes that are being brought about by the intersection of new technologies, especially the Internet and mobile technology. Although the changes brought about by earlier telecommunications technologies have not always been properly recognized or researched (he cites Carolyn Marvin's discussion of this in her 1988 book *When Old Technologies Were New: Thinking about Electronic Communications in the Nineteenth Century*), Wellman attempts to unravel how home life, work, and other social bonds are being changed today by what he calls a confluence of "technological affordances." The combination of personalized networking, knowledge-management tools, agent technology, ubiquitous networks at home and work, and the capacity to deliver almost infinite numbers of bits (both video and audio) means that society is moving from what he calls "little boxes" to a "networked individualism." (Wellman 2002). Though Wellman has produced many dozens of articles and edited collections on this topic over the past decade, I will use one of his earlier pieces (published in 2000) to convey his view.

He argues there that in an earlier phase of society,

> community has traditionally been based in agricultural villages/towns, itinerant bands, and urban neighbourhoods. People walked to visit each other in spatially compact and densely knit communities. These communities were bounded, so that most relationships happened within their gates rather than across them. They were not necessarily immobile, but even in big cities and trading towns, much intercourse stayed within neighbourhoods. Most people in a settlement knew each other. They were limited by their footpower in whom they could contact. When they visited someone, much of the neighbourhood knew who was going to see whom. Contact was between households as much as between individuals, with the sanction—or at least the awareness—of the settlement. (Wellman 2000, paragraph 7.2)

He explains that in the twentieth century,

> contemporary communities have rarely been confined to neighbourhoods. People usually obtain support, sociability, information and a sense of belonging from those who do not live within the same neighbourhood and often, not within their own metropolitan area. Community ties have been maintained through phoning, writing, driving, railroading, transiting, and flying. Most North Americans have little interpersonal connection with their neighbourhoods; they are even less subject to the social control of a neighbourhoodgroup. (7.3)

This is still place-based connectivity, but it is community-liberated:

> Liberated (from finding the community only in neighborhoods and solidarity groups). You go somewhere to meet someone or you call somewhere—to a home or an office— to talk to someone. . . . As in door to door times, connectivity is usually household to household. (7.4)

And mobile phones are creating a particular inflection:

The current move to cellphoning affords liberation from both place and group. It suits and reinforces mobile lifestyles and physically dispersed relationships. Here high speed place to place communications affords the dispersal and fragmentation of community, high speed person to person communications goes one step further, affording the dispersal and role fragmentation of the household. (7.6)

Is this good or bad? According to Wellman, this is a fundamental shift:

The structure of relationships is moving from linking places to linking people. Where place to place contact preserves some sense of contextual sense of the places where others are located, the shift to person to person contact minimizes this. People are contacting each other in ignorance of where they are operating. And because mobile people are frequently shifting from one social network to the other at the home or the office, people are contacting each other in ignorance as to what groups they are currently involved with. Rather than being embedded in one network, person to person interactors are constantly switching between networks. (7.7)

Wellman then asks, "If connectivity [is becoming] increasingly specialized as "role to role," who except household members will worry about the whole person?" (7.12). What we have, he argues, is a deeply fragmented society where there are infinite numbers of person-to-person connections with no overarching goal that brings these interactions into something bigger—hardly anything that could be called a community (or a society) at all.

THE LITERATURE ON THE MOBILE AGE

There is a vast literature on the use of mobile devices and technologies. The early compendia include James Katz and

Mark Aakhus's 2002 *Perpetual Contact: Mobile Communication, Private Talk, Public Performance*, my own coedited (with Barry Brown and Nicola Green) *Wireless World: Interdisciplinary Perspectives on the Mobile Age* (2001), and Kristóf Nyiri's annual collections on the *Communications in the 21st Century* (for example 2003, 2005, and 2009). There are Peter Glotz, Stefan Bertschi, and Chris Locke's *Thumb Culture: The Meaning of Mobile Phones for Society* (2005), Rich Ling and Per Pederson's *Mobile Communications: Re-negotiation of the Social Sphere* (2005), Joachim Höflich and Julian Gebhart's *Mobile Communication: Perspectives and Current Research Fields* (2005) and Höflich and Maren Hartmann's *Mobile Communication in Everyday Life: An Ethnographic View* (2006). Meanwhile Mizuko Ito, Daisuke Okabe, and Misa Matsuda's *Personal, Portable, Pedestrian: Mobile Phones in Japanese Life* (2005) reports on Japanese mobile life, and Lynne Hamill and Amparo Lasen look back over half a dozen years of research and attempt a synthesis in *Mobile World: Past, Present and Future* (2006). I have added to this with a collection on SMS called *The Inside Text: Social, Cultural, and Design Perspectives on SMS* (2006). Meanwhile, there have been innumerable monographs, such as Katz's *Magic in the Air: Mobile Communication and the Transformation of Social Life* (2006) and more recently, Rich Ling's (2008) *New Tech, New Ties*. The list could go on, and I am not including the rich literature on mobile phones in the anthropological canon.

All of these texts affirm that mobiles are creating communities where the copresence and mutual monitoring of door-to-door life are being replaced by the virtual presence of persons elsewhere. This, in turn, is affecting the spaces and places in which these other persons show themselves. That this is so showed itself early on. Jukka-Pekka Puro, for example, writing in 2002, argued that the mobile phone presented a new kind of stage where the information society could be acted out. At the same time as Puro was writing, Kenneth Gergen referred to the mobile phone as having *absent presence* (2002). Whether or not it is being used, the mobile phone leads the owner to think that another party is present. This manifested itself in curiously indirect ways in relationship to fashion as a means of display (like clothing) but also related to community and changing human bonds. Leopoldina Fortunati (2001) suggested that mobiles are subject to the "pull of fashion" or become fashionable in their own right in a particular way. To explain the success of mobile phones in Italy, she argued that being connected and showing that one is connected by high levels of use ensured that mobile devices came to be a fashion statement par excellence—something that Sadie Plant agreed with in the following year (2002). Mobiles enhance the self-image of the user, she was proposing, and in the process increase the user's identity within a group—by making visible the user's *invisible community*. Those who watch mobile phone users can see how popular they are, how delightful they find their friends, and how much they are sought by the number of phone calls they receive.

Behaviors that draw attention to the popularity and social fashionability of the user were labeled as a form of social flânerie in this early literature. Using them is a kind of performance and a way of behaving that is intended to draw attention to oneself. Many papers reported the different ways that this flânerie was manifest, and some noted that people who did not accessorize or personalize their phones were treated as if they were doing negative flânerie—making a statement by not making a statement. The same holds true today, when users with old phones are mocked for having cumbersome, outdated phones and poor taste in ring tones. Flânerie was also said to have different culture faces, with several researchers suggesting that in France (particularly in Paris), mobile devices were (and continue to be) used to display uniqueness through choice of device, customization, and so forth (See Hamill and Lasen 2006).

EVIDENCE ON WHO'S TALKING TO WHOM

Central to these various arguments are ones relating to the changes in the pattern, scale and scope of person-to-person connectivity. In short, this held that people would connect with more people but less deeply; widening their contacts but weakening the bonds of relationships as they did so. This view was held early on. In 2003, colleagues and I undertook a small informal survey in the United Kingdom and Germany to confirm (or disprove) this thesis (that people were connecting to an ever-increasing, socially distributed set of people and that these persons constitute members of disconnected worlds) (Vincent and Harper 2003). What we found then has been corroborated by more recent research by others, particularly Rich Ling (2008).

What we and others have found does not fit one aspect of Wellman's idea that mobile phones lead to the creation of temporary but intensive, person-to-person relations, typically built around bonds of mutual interest. The primary value of the mobile phone is instead to support activities that sustain the social lives of people who are already in social relationships—that is, for relations that exist independently of what mobiles enable. Mobiles do not widen social connectivity or the many distinct social worlds in which people operate. People who know each other before mobile telephones are bought use the phones to communicate to each other more; they do not often use them to call others outside of their social world. Mobile calls are essentially for social purposes—for friends and family to keep up with the action. Mobile phones continue to be used as tools for *invigorating* social relations, then, not for dissolving old ones and creating myriad new ones.

Nevertheless, many of those who have looked at the impact of mobile phones highlight the peculiar and different properties of how mobile technologies invigorate social relations. How they do is different from the effects of other technologies. People like Wellman and many others believe that computer and communication technologies are different only in terms of intensity of effect, but others point out that this may not be so.

For example, some have noted that mobiles counterbalance the increasing social isolation that they think is created by the use of other media, such as interactive digital TV. Some commentators have conjectured that this value above anything else makes mobile communications uniquely appealing. As long ago as 2000, Timo Kopomaa, in his *City in Your Pocket: Birth of the Mobile Information Society*, argued that mobile communications had brought Finns together in ways that other technologies had not done. Indeed, he went further and argued that Finnish culture had shifted to be more connected than hitherto. By this he did not mean in terms of total numbers of persons connected in any grouping but between those who already had social relations. Gergen (2002) argued the same and claimed that most digital media technologies—music-playing devices, digital TV, and the Internet—had taken people away from direct social contact but that mobile communications were bringing people back together, at least in the virtual sense.

Research also shows that some want this tethering more than others do. Teenagers may want to have more contact with their friends but less contact with their parents. Rich Ling explains in *The Mobile Connection* (2004) that teenagers use mobiles to ensure that they are in contact with their friends while out of reach of their parents. They use mobile phones to express and display their identity and find that the devices enable various techniques for sustaining that identity—through the way they use text, their choice of responding or not responding to calls from friends, and so on. But they don't use the same mechanism to convey their identity to their parents, who are beyond the pale of these new expressive modes. (See also Harper 2005a, 2005b; Harper and Hamill 2005). Ito and colleagues (2005) found something similar in Japan, although with more of an anthropological focus.

The sociology of mobile phones shows that these devices are then affecting society in ways that look somewhat different from those described by Wellman. The worlds that people populate are certainly small and emphasize person-to-person connectivity, as he suggests, but there is not an apparent flooding of contacts with people in diverse social roles. The fear that myriad role-to-role connections will leave the whole individual somehow less bound to others, and hence the communities of which they are a part more fragmented seems misplaced. The suggestion that role-to-role relations are somehow impoverished—compared to previously existing patterns of relations in door-to-door and household-to-household relations—might also be misplaced. Members of these small social worlds have a considerable amount of knowledge about others in these worlds because they have a great deal in common outside their mobile contacts. The mobile phone is not replacing or substituting these prior forms of relations; it is additional to these forms.

This is not to deny that mobile devices are enabling new social practices or that they can help solve problems of time and distance. But what people do when they use their mobile devices needs to be understood not solely in terms of that contact itself but in terms of how that contact operates as one of the tools of making, sustaining, and invigorating social relations in the general, not only the particular.

When people communicate with each other by phone, they are not simply solving the problem of space (and to a lesser degree, time) as if social relations could consist of merely talk (or text). Rather, they are working at those social relations and making that contact fit into larger schemes of social practices and relations in which technologically mediated expression are only a part. Once this is recognized, the whole edifice of Wellman's view appears to collapse. What he holds to be true (that communities, individuals, and the webs of connection that constitute it are altered by technology) is better recast the other way around—that community and the people that make it are rendering the technology to fit their needs. Mobile technology isn't changing community or people; community and people are determining what mobile phone systems do.

RETHINKING THE SOCIAL IMPACT OF MOBILES

The arguments presented so far has have skirted around the question of *what is being done* when people communicate. Fashion is at stake insofar as being seen to communicate is an important tool in the repertoire of fashionability. We have seen too that a new place—a domain that includes absent others—is now populating the social worlds of public and private space. In both respects, the view of the communicating individual that seems important here implies that all that matters is how often people communicate to others and how visible they make this behavior. Social status is measured by these indices—with whom, how often, and in front of others who watch (or overhear). People are nothing more than this; nor are their communities.

Put this way, the view seems rather facile. It appears insufficiently rich. After all, community must be more than simply an aggregation of people talking to each other and showing off as they do so. So what is community? There are numerous definitions of the community in the sociological canon, but I don't want to list them here. A more salient task is to reflect on why sociologists might find it hard to link any such definitions to the properties of mobile action just mentioned—to the flânerie, the invoking of absent others to justify and celebrate social status, and the ways that these behaviors are bound to other, not observable social bonds; in a phrase to particular acts of communication.

One reason might have to do with the apparent oddness of what people do, at first glance, *with* their mobile phones. One of the odd properties of overhearing a mobile phone call is the lopsidedness of the conversation one hears. The listener can't see who is at the other end of the line. To the superficial observer and to the trained sociologist, mobile telephony does seem to support a kind of social action that is stripped of the very things that Wellman says are important—the bonds that bring people together other than words. Mobile phone calls seem to be little more than exchanges of information between people who are playing diverse roles and are wrapped up in various displays of fashion.

Yet, the residues and manifestations of more familiar and more complex intentions and purposes can be found. Let me illustrate this with some biographical examples, before I then turn to explore the kinds of behavior in question more objectively. Many years ago, I was interviewing an executive of a mobile phone company, and she explained that she kept some of the text messages that her partner had sent when they were first courting. She explained, "Well, they mean something to me." At that time, text messages were cheaper than voice calls, but one couldn't properly assume that these messages were sent because the partner was concerned about money. They were both senior staff of an international corporation, and their phones were probably paid for by their employer. So cost could not be the reason. Alternatively, text messages might have been sent since they can be dealt with more easily than voice calls if received in a meeting. This may have been true, but it does not explain why the executive kept the messages.

Consider another example. In any public place, teenagers can be seen with their mobile phones. Teenagers use the technology to bring absent persons into particular social spaces. But teenagers can also be seen sharing and showing their phones to those they are with, the fellow teenagers they are standing beside. They show text messages to each other, they brag about and display the lengthy lists of phone numbers they have stored in the virtual address books, and they let their friends play with their phones. This might be merely a manifestation of fashion, but perhaps something else is going on here—something to do with how mobile phones enable people who are already physically together to be brought closer in a *social sense* as well.

For another example: all of us will have probably seen young people exchanging their mobile phone numbers by ringing each other's phones, face to face, and then adding that number with a name to their virtual address books there and then. Are they doing this because they want to increase their social contacts and create those myriad role-to-role relations that Wellman writes about? Are they doing this because mobile phone numbers are difficult to remember? Or is it to manage social actions that might be undertaken at some later date when they are apart and that will be enabled by mobile phones? Perhaps there are things that are easier to do when they are apart than when they are face to face. I am thinking here of dating and the ways in which people seek to preserve their dignity when faced with the possibility of rejection from one they want to go out with.

And for a last example, many people send and receive "good night" text messages. These are sent between couples and close friends. Do the needs being satisfied here have to do with role-to-role relations where mutual interests are strictly limited? In this regard, is it a merely a courtesy that reflects certain types of social relations and role-to-role connections? Or are these practices a hint that deeper social processes—ritual ones perhaps, related to types of intimacy—are in operation? Aren't such processes meant to die out as we move from door-to-door communities to new spaceless forms?

UNPACKING THE EVIDENCE: FROM FACE TO FACE TO TEXT TO TEXT

Why do teenagers use the mobile phones as devices to share, show, and play with when they are together? Here is an example of this kind of behavior taken from research conducted by my colleagues and myself.

Two girls, Alison and Clare, are seated beside one another in the school canteen. Alison has just received a text message and shows it to Clare and four others at the table:

Clare (speaking to Alison): I had the same one [text message]. There was this other one I was going to send you. It was quite funny, but I didn't.

The discussion continues but is inaudible. Both girls lean over the phone and talk about the content of the message:

Alison: . . . coming to you . . .

Alison tells Clare that she is sending her the message she has just received.

Alison: Clare, . . . right now!

Clare acknowledges receipt and then looks at her phone's display. A discussion follows about a message that Clare wants to send to Alison (most of it is inaudible):

Clare: . . . tell me if you get it. Okay, . . . I'm just going to show it to you because it's not going to send.

Clare is having problems sending the message. Alison and Clare start to show each other various messages they have received. Two other girls at the table are also shown the messages.

Through sharing and exchanging their phones and phone content in this way, all the girls in this context are "doing friendship." That is a sociological way of putting things, but this lies at the heart of the behaviors in question. When friendship is undertaken, one shows things—not to just anyone but to one's friends. Doing so makes those people into friends. Bound up with this is the ability of the one doing the showing to know (more or less) what things that will interest those others. Knowing what will interest them is a demonstration of how good a friend they are (or could be). Good friends make a point of knowing what their friends like. By the same token, people listen to what their friends say and look at what they are shown because they want to be interested in what their friends are interested in, too. They are friends, after all. Part of being a friend entails constituting a world in common.

Ultimately, these friendship routines produce *mutual dependence*—another soapy sociological phrase but one that is useful here since it labels an important characteristic. As one of the girls quoted above put it in a later interview, "It's all a trust thing, really. It's nice to have that with someone 'cause you don't have to say it. It's just an underlying agreement . . . that you can share."

Another aspect to securing this trust is that it has to be worked at. One cannot just show a text message as you might send a Christmas card, just once a year, and then expect enough has been done to get respect, dependence, and affection. People have to work at relationships, and this involves doing the work of sharing, looking, listening, and commenting when you are with friends. This is what the transcript illustrates. Friendship is what people do when they are with their friends. Friendship is not a merely a category or a label for a relationship type; it's a doing.

GIFTS

Some activities that people undertake when they are together are more systematic than merely showing and sharing. They involve the giving and receiving of things—in this case, the things can be text messages. The example of an executive who saved text messages sent to her by her partner illustrates a kind of *gifting*. Sharing, looking, and giving might occur face to face, but giving over distance is now made possible through the use of mobile technologies. The executive's partner sent her things from afar, and she received them. Once she received the things sent, their value (to her) was so great that she chose to keep them. These things had value for her and her memories and were constitutive of her sentiment. If these texts were ever to be shown, their intimacy would be such that only friends of special closeness would be given the honor.

In early research that Alex Taylor and I undertook on teenagers' use of mobiles, we found a great deal of evidence that they treated texts and texting this way (Berg, Taylor, and Harper 2002; Taylor and Harper 2003). The sending and receiving of text messages seemed to be a form of gift giving, including all that implies about the social patterns that ensue. The exchange of gifts is a common part of everyday life, and somehow it ties people together. Most of us take for granted that the exchange of the physical is designed to signify feelings such as thanks, caring, love, and trust and is meant to result in pleasure or well-being for the recipient. The gift, as Helmuth Berking puts it in his *Sociology of Giving* (1999, 9), "makes feelings concrete." A gift somehow embodies something of ourselves. It makes tangible something about our relationship with the one we are giving to. Giving can also help us order our memories by converting those memories into things that can be "grasped and held" and thus becomes associated with "particular histories and bound up with particular individuals" (5).

Nearly a century ago, the French anthropologist Marcel Mauss, who was probably the first to write extensively on gifts, wrote of Melanesian economies of gift exchange. But he could just as well have been talking about the text messages exchanged between young people today: "Each one, at least the dearest and most sought after . . . has its name, a personality, a history, and even a tale attached to it" (Mauss 1997, 24).

Here is a transcript of teenagers who are talking about these sorts of matters:

Jennifer: Plus you can read them [text messages] as well later. Like I can keep them and read them later.
Alex: Why do you want to read them later?
Jennifer: I don't know—if it's a nice message or something.
Susan: Yeah, Peter sends me loads of nice messages, and I want to keep them all. It's so sad 'cause he sends me so many nice ones, and I have to delete some. I feel horrible.
Jennifer: I know, and then you feel really sad.
Susan: And like I really don't want to give the phone back because it's got so many little memories and things on. And it's not the same having them written down, so I'm sending them to my other phone.
Alex: Why is it not the same?
Susan: I don't know. I know it sounds stupid but . . .

Jennifer: They don't look the same.
Susan: It's just not the same cause it's not from him anymore. It's just like me writing it down. It's just really sad. Maybe I'm just overemotional about my text messages.
Jennifer: And it's even the same when you put them in the outbox, and they lose all the time, and they lose whose it's from and everything.
Susan: That's why I think we should have memory cards because I would buy millions, really I would. . . . I really hate deleting messages that are nice, you know. Like when someone's said something that's really sweet or just like really personal or something.

Here, Jennifer and Susan are explaining how text messages and memories can become intimately entwined. Their memories are embodied in the text messages that have significance to them. They explain that texts can be used to recall past thoughts and feelings through later readings. Jennifer and Susan's remarks can help explain the example above, too, of the woman who saved texts from her partner. For her, his text messages can now be seen to bear the hallmarks of a crafted gift; made with tenderness by a special artisan: her lover. It is no wonder that she did not want to delete them.

RITUAL

Social relations are more complex and have more diverse aspects and elements than would appear to be recognized or even alluded to by those who claim that society is losing its community bonds. Community is made up of various threads, strings, and dynamics, and its various systems of exchange are not all equal or similar but bring us together into communities of sorts. And these threads, bonds, and givings don't only operate in real time, in the here and now, but tie people through time so that events in the past are brought to bear in the present and events that happen now can be marked and kept for the future.

Another aspect of how community is formed is the making of things into objects or processes that have special value. Some things are more than and distinct from the (precious) objects of memory. Their value is somehow special. They are made *sacred*. This is achieved through what sociologists label ritual.

In one of the examples I cited above, people send goodnight text messages to each other. Here is transcript evidence about this apparently ordinary but socially consequential activity:

Alex: What about you, Mark? What do you use your phone for?
Mark: Well, I mostly ring the lady [laughs] . . . and spend about half an hour. That's why my phone bill's so high.
Alex: What, talking?
Mark: Yeah, talking. Of course, I have to text her, you know, when I go to bed . . . [sounds of acknowledgment from others].
Alex: You have to? What do you mean, you have to?
Helen: It's your duty, really.
Mark: Yeah, you have to.
Susan: It's the rules!
Alex: The rules! What are the rules?
Helen: You need to say, "Good night."
Mark: Yeah, you need to say, "Good night," you need to say, "Good morning," . . .
Alex: Otherwise?

Susan: Otherwise, they get stroppy, and they dump you for being insensitive! [group laugh]
Alex: What happened before mobiles?
Helen: Well, you could phone and say, "'Night. I love you. Bye."
Mark: Yeah, I used to ring her before I went to bed. Yeah, but in the morning, that couldn't happen. Really, this [picks up his mobile] has made my life hell!

Here Mark is explaining actions that illustrate ritual. One can get into rather arcane squabbles about defining ritual, but this transcript shows that goodnight messages can be seen as transforming (albeit in a small way) the participants themselves. The messaging event has a significance to the participants that places it above mere communications. It is not the saying of good night or the receiving of text messages that is special. The doing of the entire social action gives the participants a sense of something greater than themselves — an aspiration, perhaps, that their relationship goes beyond the confines of time and space (she is there, I am here).

Although relationships between teenagers are often fragile, the accounts given by Mark, Helen, and Susan indicate how social relations are based on various aspects of reciprocity that are meant to be persistent through time. Being special for these individuals is not entirely a question of now. It has to last. Mark and his girlfriend may eventually break up, but during their courtship such ritual givings are part of what is necessary to keep it going.

This form of binding is common and old, even if the use of texting as a means to do it is not. Alvin W. Gouldner, in his *For Sociology: Renewal and Critique in Sociology Today* (1973), reflected on this kind of pattern in another context, citing the anthropologist Bronislaw Malinowski's seminal work published during the First World War on the exchange between inland communities and fishing villages (see Michael Young, *The Ethnography of Malinowski*, 1978). Using Malinowski's words, Gouldner (1973, 240) writes: "Neither partner can refuse, neither may stint, and neither should delay." Mark may or may not be gratified that he is behaving like a Trobriand islander, although he knows that the bottom line here is to keep things OK with his girlfriend. In sociological terms, Mark's texts are "a concrete and special mechanism . . . in the maintenance of [a] stable social system" (1973, 240)—in this case, with his girlfriend, a community of two.

THE FRAGILITY OF SOCIAL RELATIONS

All of this leads to the issue of not receiving messages—to the disrupting of a community. Consider the following transcript:

Jennifer: Maybe you're like, "Oh, I really want a message. I really, really want a message."
Alice: Oh, there are some days when my phone does not beep at all. I'm like, "OK, nobody likes me. NOBODY knows me!" . . .
Alex: So in a way you're—if you don't get a call or a text message?
Jennifer: You feel a bit depressed.
Alice: Because there is not a day that my phone does not go off ringing, ringing, ringing, or text messages just come flooding in. So if there's a day where it's quiet—all I get is probably one message all morning or all afternoon—I'm like, "What is wrong with the world?"

Jennifer: You think, "Have I upset someone?" 'Cause I was like that last week 'cause I fell out with somebody. I thought, "Oh, my God, maybe she's turned everybody against me," because nobody phoned me that night either. So I was phoning people—"Hello! Hello!"—having a little chat with them.

Both girls expect their friends to continue with the cycle of gift and countergift. When they fear that the cycle is broken, they feel that something is amiss—something is "wrong with the world"—and that their peers must have turned against them. Recounting one instance, Jennifer suggests that attempts can be made to reinstate the cycle through further offerings. Thus the message of "Hello!" is sent to reestablish the bonds of allegiance. The need to make such offerings alludes to the serious business of managing social relations with the delicacy they deserve. It might simply be a hello, but a great deal turns on it. Such offerings may seem mundane, but they are key to sustaining a sense of belonging, being part of the gang, being popular, and simply being OK.

These patterns of exchange are mediated through phone use (but presumably through other technologies and devices as well) and depend on reciprocity. Mobile devices provide a means of both demonstrating and testing reciprocity in relationships. The mutual dependence that derives from obligations (such as replying to text messages) binds people together, establishing and reinforcing what might be seen as the moral order of friendship and social intimacy—the bonds of a community. In sum, this sociological research shows that teenagers (and more generally, all those who have widely adopted mobile devices) use the technology to bring their social group together—not to fragment it as some sociological commentators less familiar with the evidence seem to suggest. In so doing, mobile phones are but a modern means to doing a well established practice of traditional community. As Berking (1999, 19) puts it, there is need, in any social group, to "celebrate [with] periodic consolidations of the collective in question, reproduce and evoke the requisite feelings, and thereby, in a kind of analogy with the annual cycle of offerings in archaic society, renew the foundation of the community, the normative expectations of its members, and the moral ties between each individual."

COWARDICE

In addition to new, interestingly rich practices of exchange that relate to the giving of mobile phone numbers in face-to-face situations, some things are simply easier for people to do when they are away from each other than when they are together. Mobile devices do not simply allow people to come together in the virtual sense when they are physically apart. They also allow people to do things apart that are difficult to do when they are really face to face. In allowing this, mobile phones can bring together members of communities in a social or moral sense by taking them apart in a physical one.

Consider the following example of teenagers who are talking about breaking up and asking each other out:

Jennifer: That is the worst way. That is like a bitchy thing to do. [laughs with Jenny]
Alice: That is worse than writing a letter or saying . . .
Jennifer: Our friend—he's really gullible. He's really weak, OK? He's really lovely. His girlfriend has dumped him four times in four weeks, and it was over text messages.

Alex: So why is that the worst way?
Jennifer: 'Cause it's so impersonal. It's like over the Internet as well.
Alice: It's—It's very—It's worse than being a coward. It's worse than calling someone when you know they're a thousand miles away and going, "Oh, yeah, by the way, you're dumped." It's terrible. I mean, I knew someone in the previous year—the year above me. . . . She was going out with this guy. . . . Well, she was really into him, and she thought he felt the same way and everything, so they slept together. The next day, he sends her a text. She was really, really happy, and then I think something happened at college, and then she wasn't feeling so good anyway. And then all she gets is a text. . . . You know, the happy "Oh, yeah, my boyfriend's sent me a text. He'll makes things better" thing. And it said, "You're dumped." Okay, that's not funny. And she was just crying.
Jennifer: It's the same if people ask you out over text messages. It's just cowardly.

Using mobile technology would appear to be cheap not just in the economic sense but also in the social sense. Something about using texts to dump someone or to ask them out is thought to be cowardly. We all know how hard it is to ask someone out and how hard it can be to break up with someone. But what is being alluded to here is how the use of the technology can reduce these difficulties—not because people can do these things when circumstances force them to be apart but because people can escape some of the awkward social consequences that arise when they ask out or dump someone face to face. Both social activities involve discomfort during the saying and during the response that ensues. When someone is asked out, for example, only half of the transaction is complete: the acceptance or the rejection concludes the event. Acceptance is a delight, and smiles are shared, but what happens when rejection is the response? What does the asker say at that point? How does the rejecter make the situation more comfortable for the asker?

In face-to-face situations, these social events can be difficult to conclude because those involved cannot extricate themselves from the conversation. Somehow they are obliged to take turns. One who declines the offer to go out cannot simply say no and then remain silent. He or she has to say no graciously and provide a reason: "I am sorry. I am busy that night, but thank you very much" is the kind of phrase that comes to mind. But such strategies also leave opportunities for the person asking to seek a new opportunity, such as, "Oh well, what about another night?" If this continues, the person who is trying to avoid saying no outright may well have to confront the need to do so directly and say, "Look, I don't want to go out with you." It might seem straightforward to do so, but most people put effort into not saying things that are bluntly unkind.

Social relations (even with those one does not want to go out with) are constrained by the need to be gracious, show finesse, be cool, and avoid being callous. After all, at some time in the future, the person doing the rejecting may be the one doing the asking. When people who are trying to date turn to the mobile phone, they can avoid the awkward next turn in the conversation. The phone facilitates social cohesion by letting people be *apart*.

Some traditional aspects of social relations—personal and emotional aspects—can cause anguish, in other words. Digital mediation allows people to minimize these problems, smoothing their efforts to bond emotionally. In this view, technology is not rendering communities asunder but enabling members of those communities to avoid pain and thus keep their communities together. It does so precisely in the opposite way that one might imagine communication technologies might do, by keeping *them* apart—the people, with all their passion, fear, and hope. Sometimes these are too much to handle when people are together.

CONCLUSIONS: MAKING TECHNOLOGY FIT SOCIETY

The empirical goal that I have for this chapter is twofold. First to show that the bonds people create between each other are not being rent apart by new communications technologies—at least not in the sense that many sociologists feared it would when mobiles first became widespread. People continue to create bonds with other people and mobile telephones turn out to offer particular advantages for this.

A second goal has been to show how one particular communications technology, mobile phones, do this in particular ways. What we find is that these ways are at once diverse, startling and subtle, and at the same time often similar to and driven by modalities of human bonding that have profound and long historical roots.

These ways do not complement each other in neat and tidy ways, as if the bonds that make up the weave of society are neatly fitted together in everyday action. The texture of these bonds, what is produced by acts of communication, is rough in some places, smooth in others, weak elsewhere. It is constituted in ragged and diverse ways, whilst others remain the same. Some are wholly novel, some pretty much as they have been for years, even centuries. Gifting is hardly a twenty-first-century phenomenon, even if the material given is contemporary.

This does not mean that the acts in question or the texture of bonds they create cannot be understood or mapped, but subtlety needs to be applied when analyzing them. As has been shown, keeping a lover's text message is an instance of a mobile enabled act of communication. Although a text is small in terms of digital content, its social value as a tool for making a bond between those involved should not be underestimated.

More broadly, one should also learn the ways that new technology gets used to make bonds has a vitality to it—sometimes leading to innovative in those practices, for example when the ritual delicacies of dating are smoothed through remote messaging, and sometimes creating pale shadows of past ways. A SMS love note is doubtless less evocative as one written in pen and ink (See Höflich and Gebhardt's paper, "Changing Cultures of Written Communication," 2005). Mobile devices are used to sustain and create systems of reciprocity, obligation, and the simple daily construction of trust and intimacy between friends. These devices are not merely end points in a system of peer-to-peer communications. The bonds that are created by them are more complex than this.

Because of the vigorous way that people adopt them, mobile devices have allowed certain changes to manifest themselves in the arrangements constitutive of community. What is changing is not what many commentators suggested at first. It is now not thought that communities are less cohesive than they were, although the ways that various communities achieve solidarity are different. As noted, the giving and receiving of text gifts is a new practice, but it harks back to many prior forms of gifts and exchange. Using mobile technology to mediate the painful work of creating and ending emotional relationships is new, too, although it does not reduce pain but simply makes it easier to deal with. And sharing and showing what one has stored on one's mobile device are new ways to build friendship. In these apparently small ways, communities are changing while they remain the same.

I have not listed all the ways that mobile devices are adopted for the purposes of social bonding. Producing an entire taxonomy is not my goal here. Any list of possible purposes for mobile devices should support the claim that mobiles are being made to fit what people need to do (rather than the claim that what people do is being made to fit what mobiles enable). Although the examples I cite relate primarily to text and voice messaging (the primary set of functionalities of mobile technologies over the past decade or so), other communication media will afford their own ways of being shaped and in creating different modalities of being in touch.

A sociological view uses a framing mechanism to highlight certain sorts of issues and topics, and in the case of mobile devices, sociologists assumed early on that community would weaken its moral fabrics and shift toward more anodyne forms. Much of my evidence is from a micro sociological level and thus might not pertain to large-scale trends affecting community. But we have seen that communicative practices need to be understood in their details. These have been the topic of this chapter. The suggestion that sociology can be neglectful and that this might be a natural consequence of its way of looking at human behavior cannot be dismissed altogether. Even before he wrote the paper on spaceless communities mentioned above, Wellman had noted that Internet studies seemed to take two opposing points of view that allow little room for more subtle appraisals that would fit somewhere in between (Wellman and Gulia 1999, 13):

> Manicheans on either side of this debate assert that the Internet will either create wonderful new forms of community or will destroy community altogether. These dueling dualists feed off each other, using unequivocal assertions of the other side as foils for their own arguments. Their statements of enthusiasm or criticism leave little room for moderate, mixed situations that may be the reality.

Thus, Wellman himself appears to be receptive to the claim that his view of social change—a society made up of individuals whose bonds have moved from door to door to role to role—is itself an idealized vision that needs to be grounded in more thoughtful explorations of evidence.

What is interesting about these discussions of the sociology of mobile phones is not that it turned out that mobiles don't have the impact that many fear. What sociologists have been in the business of doing (irrespective of whether they have good evidence for doing it) is rebuking ordinary people for not attending to the serious threats to society that technology is introducing. Sociologists often claim that people are innocently slashing society into pieces by using new and more widespread transport systems, the Internet, or their mobile devices without regard for the consequences. Yet the evidence suggests that for the mobile phone (the most ubiquitous of mass computing devices), almost the reverse is true. The phone seems to be sustaining and invigorating social relations. Many people seem to expend great effort on their use of mobiles. Part of their seriousness comes about because they worry not about society coming apart (as is argued by Putnam, for example, in his *Bowling Alone*), but about the fragility of their friendships and their families—their emotional partnerships. These are constitutive elements of society even if they are not its totality. The people in question know that their efforts have meaning in their lives. Failure can hurt both themselves and others in their world.

In 1986, the anthropologists Catherine Lutz and Geoffrey White noted that emotion had systemically been avoided in most anthropology up to that time. When it was dealt with, the perspectives applied were so crude and simplistic as to be virtually worthless. They argued that anthropologists needed to capture the role that emotion played as part of the assembly of factors that constitute social experience. Only in this way could "what is at stake for people in everyday life" be understood (431). Sociological studies similarly neglected predicting how individuals use of mobile technologies in serious and emotional ways. Vital aspects of the success with which people adopt mobile phones are both the skill with which they use these devices to cement their relationships as well as the faithfulness with which they undertake those activities. People don't text each other because they are thinking about how to balance giving and receiving. They do these things practically without thinking. It comes naturally. More accurately, it comes from the heart. A person who tries to develop a relationship by counting the givings and the takings will be seen and judged for what they are—one who calculates rather than feels, who measures rather than loves, who has little merit and much ambition.

Recognition of the morally reprehensible in this new world is made possible by the ways in which technologies like mobile phones are made into tools for making connections between people, for making a texture of bonds. The strength of this texture might be under threat but not from technology. The threat comes from those who act without faith—in themselves, in others, and in the communities of which they are a part. Communicative expressiveness—artfulness in the sayings, writings, and textings enabled by technological mediation—are central to the effectiveness use of new communications tools. As I note in chapter 1, we are increasingly judged by what we express—by our communications. What this chapter has shown is the depth, diversity and importance of the communication nexus between individuals with just one technology. The shape and form of this nexus is bound up with the kinds of affairs that people seek to undertake with one another, as well as the tools they have at hand to achieve their sought for ends. Indeed, communication goes to the heart of their daily affairs.

It seems to me very likely that this will apply to their use of other technologies too—new social networking sites on the Internet, new blogging techniques (such as Twitter), and so on. It may not apply in the same way, though. If research on mobiles shows that people use this technology to bind themselves together in their emotional lives, it is certainly not true to say that some of their other modes of communication address the same elements of their lives. Although blogging may be about emotional expression, the emotions in question are not related to desire; they are related to anger, truth and indignance about issues of public concern. These are about the topics that society is thinking about, blogging about. But this turns out to be a serious business too. Bloggers do get passionate about the benefits of blogging, asserting the claim that blogging is vital to developing democracy, for example.

Other contemporary modes of connectedness are treated seriously, though in different ways. If blogging can be thought of as a way of creating the *digital commons*, the same cannot be said about many social networking sites. These are designed to let people make the kinds of bonds that mobiles allow them to. But if it was the case that the impact of mobiles was expected to be damaging, then one also hears similar fears expressed about Facebook and other social networking sites. Though commentators like Clay Shirky, in *Here Comes Everybody* (2008) emphasize the virtues of "networked identity," he focuses on the political aspect and explains that it's a sense of civic virtue that motivates this. But the personal aspect, how these sites might affect the management of affairs of the heart, friendship and family life, are things that many do worry about. There is a concern that the bonds might be ill-advised, with young girls being particularly vulnerable to older males for example. There is a fear too that the boundaries between the private and the public will blur, and that this might affect people's prospects in the future. A prospective partner might be put off by embarrassing and shaming images.

But that there are such concerns may not simply be related to what social networking sites might do. It might be that acts of communication of all kinds are treated as a serious matter. As we move into the second decade of the twenty-first century, we might think that we are turning into communication obsessives, but careful examination of how we bind ourselves together through acts of communication shows that we have been putting a great deal of effort into this for some time. Our efforts would appear to increase as we adopt new communications technologies. Our fears that these new technologies might undermine our efforts a reflection of the importance we give acts of communication. How strange, then, that sociologists might warn us to lament our failures to maintain community through technologies of communication. How blind and divorced from what people really do. How out of touch they have become in a world where being in touch is perhaps more serious than ever.

REFERENCES

Berg, S., A. Taylor, and R. Harper. 2002. Mobile phones for the next generation: Device designs for teenagers. In *Proceedings of CHI 2003* (433–440). New York: ACM Press.

Berking, H. 1999. *Sociology of Giving*. London: Sage.

Brown, B., N. Green, and R. Harper, eds. 2001. *Wireless World: Interdisciplinary Perspectives on the Mobile Age*. Heidelberg: Springer Verlag.

Castells, M. 1996. *The Rise of Network Society*. Malden: Blackwell.

Castells, M. 2009. *Communication Power*. Oxford: Oxford University Press.

Clark, A. 2003. *Natural-Born Cyborgs: Minds, Technologies and the Future of Human*. Buckingham, U.K.: Open University Press.

Fortunati, L. 2001. The mobile phone: An identity on the move. *Personal and Ubiquitous Computing* 5(2): 85–98.

Gergen, K. 2002. The challenge of absent presence. In J. Katz and M. Aakhus, eds., *Perpetual Contact: Mobile Communication, Private Talk, Public Performance* (227–241). New York: Cambridge University Press.

Glotz, P., S. Bertschi, and C. Locke, eds. 2005. *Thumb Culture: The Meaning of Mobile Phones for Society*. Bielefield, Germany: Transcript Verlag.

Goffman, E. 1959. *The Presentation of Self in Everyday Life*. New York: Anchor.

Gouldner, A. W. 1973. *For Sociology: Renewal and Critique in Sociology Today*. London: Allen Lane.

Hamill, L., and A. Lasen, eds. 2006. *Mobile World: Past, Present and Future*. Godalming: Springer-Verlag.

Harper, R. 2005a. The moral order of text: Explorations in the social performance of SMS. In J. Höflich and J. Gebhard, eds., *Mobile Communication—Perspectives and Current Research Fields* (199–222). Berlin: Peter Lang GmbH—Europäischer Verlag der Wissenschaften.

Harper, R. 2005b. To teenage life to Victorian morals and back: Technological change and teenage life. In P. Glotz, S. Bertschi, and C. Locke, eds., *Thumb Culture: The Meaning of Mobile Phones for Society* (101–113). Bielefeld, Germany: Transcript Verlag.

Harper, R., and L. Hamill. 2005. Kids will be kids: The role of mobiles in teenage life. In L. Hamill and A. Lasen, eds., *Mobile World: Past, Present and Future* (61–73). Godalming: Springer-Verlag.

Harper, R., L. Palen, and A. Taylor. 2006. *The Inside Text: Social, Cultural, and Design Perspectives on SMS*. Dordrecht: Springer.

Höflich, J., and J. Gebhart, eds. 2005. *Mobile Communication: Perspectives and Current Research Fields*. Berlin: Peter Lang GmbH–Europäischer Verlag der Wissenschaften.

Höflich, J., and J. Gebhardt. 2005. Changing Cultures of Written Communication: Letter-Email-SMS. In R. Harper, L. Palen, and A. Taylor, eds., *The Inside Text: Social, Cultural and Design Perspectives on SMS* (9-32). Dordrecht: Springer.

Höflich, J., and M. Hartmann, eds. 2006. *Mobile Communication in Everyday Life: An Ethnographic View*. Berlin: Frank and Timme.

Ito, M., D. Okabe, and M. Matsuda, eds. 2005. *Personal, Portable, Pedestrian: Mobile Phones in Japanese Life*. Cambridge, MA: MIT Press.

Katz, J. 2006. *Magic in the Air: Mobile Communication and the Transformation of Social Life*. New Brunswick, NJ: Transaction.

Katz, J., and M. Aakhus, eds. 2002. *Perpetual Contact: Mobile Communication, Private Talk, Public Performance*. New York: Cambridge University Press.

Kopomaa, T. 2000. *The City in Your Pocket: Birth of the Mobile Information Society*. Helsinki: Gaudeamus.

Ling, R. 2004. *The Mobile Connection: The Cell Phone's Impact on Society*. San Francisco: Morgan Kaufmann.

Ling, R. 2008. *New Tech, New Ties: How Mobile Communication is Reshaping Social Cohesion*. Cambridge, MA: MIT Press.

Ling, R., and P. E. Pederson. 2005. *Mobile Communications: Re-negotiation of the Social Sphere*. London: Springer.

Lutz, C., and G. M. White. 1986. The anthropology of emotions. *Annual Review of Anthropology* 15: 405–436.

Marvin, C. 1988. *When Old Technologies Were New: Thinking about Electronic Communications in the Nineteenth Century*. Oxford: Oxford University Press.

Mauss, M. 1997. *The Gift: The Form and Reason for Exchange in Archaic Societies*. London: Routledge.

Nyiri, K., ed. 2003. *Mobile Democracy: Essays on Society, Self and Politics*. Vienna: Passagen Verlag.

Nyiri, K., ed. 2005. *Communications for the Twenty-first Century*. Vienna: Passengen Verlag.

Nyiri, K., ed. 2009. *Engagement and Exposure*. Vienna: Passengen Verlag.

Plant, S. 2002. *On the Mobile: The Effects of Mobile Telephones on Social and Individual Life.* Report commissioned by Motorola, http://www.motorola.com/mot/doc/0/234_MotDoc.pdf.

Putnam, R. 2000. *Bowling Alone: America's Declining Social Capital.* New York: Simon and Schuster.

Puro, J.-P. 2002. Finland: A mobile culture. In J. Katz and M. Aakhus, eds., *Perpetual Contact: Mobile Communication, Private Talk, Public Performance* (19–29). New York: Cambridge University Press.

Rheingold, H. 2003. *Smart Mobs: The Next Social Revolution.* Cambridge: Perseus. Available at http://www.umts-forum.org/servlet/dycon/ztumts/umts/Live/en/umts/Resources_Reports_26_index

Shirky, C. 2008. *Here Comes Everybody: The Power of Organizing without Communications.* London: Allen Lane.

Taylor, A., and R. Harper. 2001. Age-old practices in the New World: A study of gift-giving between teenage mobile phone users. In *CHI 2002* (439–446). Minneapolis: ACM Press.

Taylor, A., and R. Harper. 2003. The gift of the gab: A design-oriented sociology of young people's use of mobiles. *CSCW: An International Journal* 12 (1):267–296.

Thrift, N. 1996. *Spatial Formations.* London: Sage.

Thrift, N. 2004. Movement-space: The changing domain of thinking resulting from the development of new kinds of spatial awareness. *Economy and Society* 33: 582–604.

Urry, J. 2007. *Mobilities.* Cambridge: Polity Press.

Vincent, J., and R. Harper. 2003. *The social shaping of UMTS,* available at http://www.umts-forum.org/servlet/dycon/ztumts/umts/Lilve/en/umts/Resources_Reports_26_index.

Wellman, B. 2000. Changing connectivity: A future history of Y2.03K. *Sociological Research Online* 4(4). Available at http://www.socresonline.org.uk/4/4wellman.html.

Wellman, B. 2002. Designing the Internet for a networked society: Little boxes, glocalization, and networked individualism. *Communications of the ACM* 45:91–96.

Wellman, B., and M. Gulia. 1999. Net surfers don't ride alone: Virtual community as community. A preliminary version is available at http://www.acm.org/ccp/refercnes/wellman.htm.

Young, M., ed. 1978. *The Ethnography of Malinowski: The Trobriand Islands, 1915–18*. London: Routledge and Kegan Paul.

5 SOMETHING TO TELL

PREAMBLE

Our age is being defined by our relationships with devices that were initially designed simply to calculate—computers. Yet computers now come in all sorts of shapes and sizes, and their ability to calculate binary bits has become secondary to other things that they appear to do for us. A digital camera is a computer of sorts, but when it takes pictures, we don't think of it as taking measures of light within a matrix of evenly distributed light-sensitive zones. We want the camera to let us show pictures, not indexes of light, to other people. Similarly, our mobile phones let us communicate our voices, not binary representations of radio wave frequencies. But saying that computers have become virtually invisible misses the point—which is that our relationships with computers have come to stand proxy for parts of our relationship with people. Computers compute and thus shape our economic world, but they also connect us to others, thus shaping our social world.

Over the past twenty-five years or so, the evolution of computing—from mainframe to mobile, from desktop to wearable, from letting us do desktop publishing to letting us tweet—has reflected and been reflected in this extension of function. These changes have to do with what we think we are, what we do, and what we think the devices that saturate our lives will let us be. Sherry Turkle, in her book *The Second Self: Computers and the Human Spirit* (1984), written when the first personal computers were beginning to be assembled in basements and backrooms, suggested that at that time we saw ourselves mirrored in the output of the code we had written in the machines. She noted that users (pioneering hackers) programmed the devices in front of them and that what those computers did—indeed, all they could do—was a reflection of the coder. Those who programmed felt as if their minds had been imposed on the computer and that as a result of their instructions computers reflected their imagination. Hence her book's title: computers were the second self of the programmer, reflected in silicon.

Turkle was not alone in seeing the links between people and computers—between the intimate goals of the hacker and the early PC. Educational theorists, for example, argued (for a while, at least) that programming and programming languages could be the task and the tool that would transform the inner workings of school kids' minds. Programming skills required abstract reason, and it also would let kids assert themselves in ways that had never before been possible—or even imagined. With PCs and the possibility of coding in the classroom, children who had merely read books and absorbed their contents could now create content with new narratives and new structures. Like Turkle's programmers, they could produce a second self, and the educationists had in mind a better self—an educated, abstract-reasoning, inner self (see Tom Conlon's 1985 *Start Problem-Solving with Prolog*). This also parallels arguments from literary theory and hypertextuality that emerged somewhat later in the work of Jay David Bolter (1991) and George P. Landow (1992), among many others.

All this now seems quaint. The landscape has changed greatly since the 1980s. As we look back, we can see how little our inner selves seemed to change when we learned computer science. We can see too how feeble and impoverished those digital second selves were. They struggled to undertake even the simplest word processing tasks. Overheating and breaking, they sometimes took ten minutes just to load up the operating system (OS). Perhaps we forget the intoxicating magic of having a machine do as we command—as we program it to do.

TODAY'S COMPUTER-HUMAN LANDSCAPE

The landscape has changed in ways that are fundamental, even if those changes aren't all technological. Computers now seem to serve human communication more than they do almost anything else, even shape our economic lives. The character and salience of the person-machine nexus that is precious in Turkle's book is altogether different if not dissolved. One of the most visible reasons for this change is the graphical user interface, which has transformed the relationship that people have with computers, especially PCs. When Turkle was doing her research, hackers were toying with the idea of using colorful objects on a screen to represent what the computer was doing, but they were primarily interested in seeing lines of code (lines that they had written) on a screen. Today one has difficulty finding any lines of code at all on a computer screen, except perhaps for the uniform resource locator (URL) of Web sites we are visiting. All the stuff that so intoxicated those pioneers—the algorithms that commanded computers—have been replaced by graphical objects that represent not the code but the tasks that the code supports. Similarly, word processing today entails not selecting an appropriate command line instruction but selecting a pull-down menu with lots of options either in plain English or in symbols like signs on a motorway. In these ways, that a computer might be a second self has been lost from view, submerged in the icons and tools that have somehow made that computer hide itself so that we can focus on other things—such as writing emails and composing Web blogs.

Another important change has had effects underneath the hood of the PC—the widespread adoption of PCs with a TCP-IP communications stack inside. For Microsoft, this stack has been crucial for changing people's perceptions of what a PC can do, since it was part of all Windows 95 releases and allowed people (who had a telephone and a modem) to link one computer to another easily. Windows 95 was the first operating system that was purchased by enormous numbers of people, and it introduced millions to computer communications. This change was linked to the simultaneous invention of the graphical browser (most notably Netscape and then our own Explorer), as well as the hypertext markup language (HTML) standards that produced content for such applications.

These and other changes have altered forever the relationships people have with computers. Not surprisingly, given their range and consequence, there is no common agreement as to what they mean. They allow communication to others, for example. Turkle offers her own view in a revised edition of *The Second Self: Computers and the Human Spirit, Twentieth Anniversary Edition* (2005), where she says that twenty-five years ago the computer acted as a mirror of the programmer (it was reflective) and that now it fosters simulation. People have computers so that they can do things with them. They're seeking not to see themselves reflected but to imagine other selves, other ways of being. Turkle might have been perplexed by the second selves that were being conjured in the early 1980s, but in 2005 she suggests that we were all being delighted by the fictional second lives we are creating in the virtual world of cyberspace (hence her shift from the *Second Self* to the Second Life). Not everyone agrees with this view. As Clay Shirky has pointed out in *Here Comes Everybody: The Power of Organizing without Organizations* (2008), one's presence on the Web can be fictional, but in practice there has been a strong assertion of social values that emphasize the need for accuracy and truth. One has to be who one is, even in the digital ether, and this has been the case on the earliest social networks sites (like LamdaMOO) and on the more recent ones (like Facebook and MySpace). Research studies of new communications technologies have shown that they get used in largely earnest ways. Who you are turns out to be something that one can play with—but not too much.[1]

WHY COMMUNICATE?

Understanding our relationships with computers is central to this chapter. But determining what that relationship is requires some deliberation. As I remarked earlier, whatever this relationship might have been in the past, today it stands proxy for another set of relationships—ones that we have with people. But why do people have such a strong desire for using computers as communications technologies—as means for being in touch (somehow) with other people? The social context of this use is looked at in the prior chapter, which explores the relationship between people and society as mediated by technologies. Research shows that society is not being transformed in quite the way that some commentators predicted and that communications technologies can sometimes bring people together and at other times push them apart. Technologies designed for remote communication can sometimes deepen the sense that people have of being together even when they actually are together. The effects of communications technologies are curious indeed.

These effects are driven by what we saw were powerful motivations, many of which are very old: saving face is a time-honored problem just as is giving gifts. Is it possible to define the charms of technologically enabled communications that fit in-between these extremes? What is it that sometimes entices people to delight in a channel and at other times pushes them away? From what has been discussed thus far we wouldn't be able to answer why some people engage indulgently with technologically mediated communication, for example. I use *indulgently* here to allude to the possibility that people are conscious of what they do and fret about how to tame their communicative urges when they communicate indulgently. This mindfulness is what I am after—consciousness of the communicative act when the consciousness appeals to matters that are bound up with the passions and revulsion that expression—that communication—can induce.

As I describe in chapter 3, the creative landscape in which I have operated for much of my career has been based on a view of the human that emphasizes bodiliness. Our research forays into communications have most often been based around geographic maps of human lookings and glancings—of telecommunicated replications of touches and pointings. The communicating human, in our view, is a body that expressed itself in movement. Recently, however, this view has been failing us. It hasn't ceased to provide fruitful ground for invention, but the concepts and ideas that it has allowed us to produce have delivered experiences—even enchantments—to and for the users that the model has not helped us understand or explain. In the past two or three years, my colleagues and I have had to search for new ways of understanding communication acts because we have been perplexed by the ways that our users have been using our technologies—the things we invented for them. We have read the founding fathers of this concern for the body (and its acts) within computer science and reflected on the limits of mathematical models of communication to answer new questions. We explored sociology, for example (and the work reported in the prior chapter), and learnt about the vitality of human bonding through acts of communication; we investigated historical studies like Henkin's, too. And it led us to start thinking about the aspects of communication acts that hitherto we had not focused on or had ignored.

I want to illustrate this evolution in our thinking by reference to two technologies of our own devising. Their use in trials with subjects here in Cambridge perplexed us. The devices in question were rather simple. They were rather commonplace in purpose and style, too. Whereabouts clocks supported text messaging and location awareness, and Glancephones allowed people to use mobile networks to glance at each other. Each device was invented as a way of fitting (at least partial) technology solutions to need. But each ended up being used (and could only be understood) by treating the communicative acts in question quite differently from how we had done so before. Communications here weren't about increasing sensory depth, for example, or providing a closer fitting of two (or more) communicating human bodies. They were about users' sensibility for communication that was fostered in part by use of the technologies themselves. These technologies did not afford a fitting. They created human expressive desire.

GLANCING

Of the two devices, Glancephones were designed more obviously around the idea of fitting, and this in turn was an elaboration of a view that emphasized the body when thinking of the communicating human. This view led to an idea that mobile phone communication should simulate certain aspects of social interaction that had hitherto been neglected or at least not made possible with mobile phones. The communication acts in question related to the structural patternings that are visible when one person says hello to another—a *greetings sequence.* At the time of the research, mobile phone technology did not allow the normal pattern of this to occur in ways described below. It was thought that offering some kind of replication of a face-to-face greetings sequence would appeal to users, making mobile phones seem more natural to use.

A greetings sequence is a fairly basic feature of everyday conversation (something that was studied most notably by Harvey Sacks in the 1960s. See Jefferson and Schegloff's edited Sacks lectures from 1992). When people seek to converse with another, they commence the conversation with a greeting, and this prompts a greeting response from the person addressed. The two stages are connected, so when a person does not reply to a greeting, it is thought to be a case of rudeness or insult. If a person says hello, the other is obliged by the rules of etiquette to say hello back.[2] For mobile phone communications, a greetings sequence could not take this form. A mobile-enabled greeting offered a distorted version of the tradition sequence: when someone called another, the other's phone rang and stood in as a surrogate or proxy for the first person saying hello. The recipient of the call could then press the phone's relevant button to say "Accept," and this would be an answer to the greeting: it would be their hello back. This might seem fairly close to normal interaction until it is dissected some more. For example, the person making the first step (the first hello) won't be able to vary the tone of the hello to reflect his or her feelings. The caller may be angry or sad, joyful or despondent, but when the other's phone rang, the hello would always be the same. A whispered hello would be the same as bellow, and a shout would be as good as a murmur. This seemed an obvious failing in mobile system design, we thought (just as it is with landlines of course).[3]

Another feature of mobile phone systems of the early 2000s created further distance between the normal and the telemediated. With mobile phones, an individual was able to choose a ring tone for a particular caller so that whenever the caller contacted that individual, that personalized ring tone would be produced by the phone. In this way, the recipient of a call could anticipate that the caller's hello was going to be shouted, whispered, or mumbled (according to their normal way of speaking). But recipients of these kinds of calls could give themselves completely wrong indications of a caller's mood. Ring tones allowed the recipient of the hello to choose the manner of the hello. In ordinary conversations, this is not possible. This aspect of mobile phone technology inverted what one might call the normal rules of communication.

It seemed to us that there were lots of ways to correct this misfit of human greetings sequences. We thought we could design improvements to mobile phone systems that would bring those systems closer to the human norm. One idea we had was for callers to choose a ring tone that would be sent to the recipient's phone to reflect the mood of the sender. This might delay the connection slightly, we thought, since the recipient's phone would have to download the ring tone and install it before it could start ringing. As we thought about this solution, we also recognized that it would create some problems (beyond the momentary delay created by the loading of the sender's mood tone). The recipient would know the mood of the caller but not the identity of the caller. On closer reflection, we thought this would easily be avoided if the caller's number was in the recipient's address book, since the caller's name would automatically appear on the phone's screen when the ring tone started to play.

Nevertheless, these reflections led us to create a stage in a greetings sequence that came *before* the first hello. In face-to-face situations, people glance at each other before saying hello (Sudnow 1972). They do so to see whether the other person is available to talk or is doing something else (reading, perhaps) that would make talking seem to be an interruption. Glancing can also reveal another person's mood, and this might affect how someone chooses to open a conversation. If the person looks sad, they might say, "Sorry to disturb you" or "You are looking downhearted"—thus prompting a reply that explains that expression.

So we opted to design an application that would allow users to *glance*. The goal for our design solution was to allow callers to glance at the person they were seeking to contact before they said hello. Having glanced, they could judge whether it was a good time to call. The glancing might also enable them to judge what might be the most appropriate opening gambit ("You look worried"). Any design, however ingenious, would not be able to replicate perfectly what human glancing allows. A phone in the pocket would not allow much glancing, for example, except into darkness. But we felt that it would be a benefit if phones could be set up (somehow) to allow glancing when it was appropriate. It would afford at least a better fit (if not a perfect one) to people's social needs.

As we reflected on these issues, our idea gradually evolved into Glancephones, which were, in essence, camera phones that could be set up so that a caller could glance through them. To achieve this required quite a bit of engineering of the hardware and software. We started by buying standard camera phones that had front-facing cameras (many camera phones have cameras facing away from the back of the phone). We chose phones that allowed this front-facing camera to be switched on by the movement of a slider on the case (not all camera phones function in this way). We attached a little leg to this slider so that when this slider was moved down, the leg moved out, making a tripod effect on the base of the phones. When a user turned on the camera by moving the sliding leg, the phones stood up and functioned like a webcam. Once in this mode, we reasoned, a glance would be possible.

Having sorted out the hardware, we then wrote an application that we installed on the phones. When the phones were put in the tripod mode, a Glancephone application started automatically, and ordinary phone calls could not be made. Only glances could be made when someone called the phone. When someone did, they were told this by a screen dialog.

If they pressed the yes soft key when prompted, a glance was undertaken. It turned out to be difficult to get a video connection at this point, so we designed the application to take a still image and send this as a *glancepacket* across the mobile networks to the glancing phone. This still image turned out to be the glance and was displayed on the caller's screen much like an MMS (Multimedia Messaging Service) picture. Since glancing in real life is often reciprocal, such that one might glance back at those who glanced, we also decided to design the application to indicate who had glanced. We cropped the name of the caller from the address book and displayed this on the phone screen of the person whose phone had been glanced at. To achieve this, we had to ensure that the callers had their name in the address book.

FINDINGS

Glancephones also had other features, but the point is that our design reflected our presumption that users would want some kind of better fit between their ordinary natural communications and the telecommunicated version. Yet despite all our reflections about normal practices and our attempts to offer a digital fit for human communicative practice, actual use of Glancephones took quite a different form.

We imagined that users would put their phone in glance mode when they were happy to be glanced at and less happy about being interrupted with a normal call—when they were in meetings, say, or at dinner with their families. We thought that recipients would rely on the glancing functionality of their phones to reduce interruption. A caller would see in the glance that a phone call would be either appropriate or intrusive, and if the latter, they would therefore delay the greeting until later. But we found that people set up their Glancephones on occasions when they *wanted* to be glanced at and when they felt that they had something that was worth seeing. This worked in the following, somewhat curious manner. A user would decide that they were doing something that their friends should glance at and see. To get their friends to glance at them, they would try and glance at them first, which notified their friend that someone wanted to be glanced at in return (see Goodwin 1979).

Within a few days of distributing the Glancephones, users came to know that an attempt to glance was an elicitation for them to glance *back*. In one instance, a user set up his Glancephone on a restaurant table and sought glances so that his friends could see the expensive wine he had bought. Someone else set up a Glancephone so that others glancing could see they were on a date with someone special. Generally speaking, glancing was sought when people thought they were doing some thing their friends would envy. In another case, a user set up a Glancephone to show that he was sitting at home watching television, knowing full well that some friends were working late. There were many similar examples.

This digital form of glancing had some of the proprieties and social consequences of glancing in real life. It was bound to judgments about who could glance and what was worth showing and seeing. But it was bound to larger narratives of playful interactions between the participants. In this and in various other respects, it was essentially a different experience from glancing as an element of greetings sequences. Glancephones were not being used to finesse the gentle rhythm of summons and answers. The devices were being used to get friends to look at oneself. This was attention getting, not glancing.

What does one learn from this research study? To begin with, our notions of natural communicative behavior provided a design rationale, but it did not lead to a system that was used in a way that reflected the ideal form of behavior we had set up. Glancephones simply did not get used in a way that we had planned. Instead, they got used to do new things, and those new things were subject to an emergent form of social etiquette—"You glanced me, so I'd better glance back at you"—a form of behavior that was only partly similar to the greeting sequences that had inspired our design. This glancing etiquette wasn't about how to deal with interruption. It was about laughter, mischief, even vanity—about a kind of performance.

If one thinks about the overall topic of the book (one aspect of which is communicative volume), it seems apparent that one can hardly start thinking about these concerns—laughter, mischief, and vanity—arithmetically. Nor can one think about reaching a point where there is a perfect technological replication of natural communication about these matters. Indeed, one wonders whether it makes sense to talk of designing a system that allows laughter, mischief, and vanity to be conveyed efficiently. Glancephones led to the recognition that new communicative media ought to be understood in terms of how technology can be deployed within social codes of behavior and conduct; a moral order if you will. This certainly affirms what one learns for the sociological studies of mobile phone used discussed in chapter 4. But usage of Glancephones shows aswell that people extend and evolve this moral order in ways that gives it surprising new forms. To use SMS as a form of gift giving is to mirror certain social actions with digital means for example; to use mobile phones as a way of saving face in awkward situations tells us how important some moral codes are, codes that will exist with or without communications technology. Similarly, one can readily acknowledge the commonplace fact of vanity that leads some people to blog. The desire of some bloggers to be listened to all the time, via Twitter, a reflection less of the technology itself than of this fact about them. Blogging fits something, allowing something to surface that needs to. Communications technology allows people to do what they have always done and always will do. But Glancephones allow glee about new doings.

This offers a further route to the question of what people mean when they communicate; it's not merely the semantics of the words they use but their purposes in doing so, too. Here the doings are those of both the one glanced at and the one doing the glancing. Those who delighted in the Glancephone liked to celebrate their lives and friendships through laughter and mockery. This entailed self-celebration (as when a person got others to glance at him) and self-deprecation (as when those doing a glance acknowledged in subsequent turns at communication that what they were glancing at was indeed more interesting than what *they* were doing). *Volume*, *capacity*, and *communicative burden* are phrases that are commonplace in communications engineering, in human-computer interaction, and other related disciplines that seek to invent for communication and are orthogonal to these matters.

My colleagues and I learned with our Glancephone research that what mattered in communication was not what we had originally thought. We came to see that Glancephones conveyed a certain picture of their users—not in terms of fixed visual representations but in terms of a certain view of the things that our study participants did. Glancephones came to be a tool to help them build a character study of themselves for their friends. As we came to recognize this, we recognized too that not everyone would find our Glancephones appealing. Not everyone would want to convey a sense of their character in their communication acts in just this way. But one important lesson from the study was recognizing that communications are about our humanness in the broadest sense—about who we might be and about how that who could be conveyed in the act of communicating.

WHERE ARE YOU?

The Whereabouts clock had the same eponymous motif as Glancephones. The latter supported glancing, and the Whereabouts clock supported knowing where people were. The device took its design style from the delightful clock in the Weasley house in the Harry Potter book series, which indicated where members of the Weasley clan were at any time. Our device also displayed the location of its users, and these locations were displayed on a circular touch-sensitive screen, with any messages they had sent hidden beneath an image of the user in question. These messages could be opened by touching the image in question. Although the Weasley clock offered detailed indications of arrival (down to the second), our own offered only rough indications that users might be at school, work, or home or in some unspecified other place. The system behind the clocks used mobile phone cell identification to generate location information and combined this with any message content sent over the mobile networks. Thus, when a message was delivered, it was displayed on a clock screen in the zone from which it was sent. Even when a message wasn't sent, an icon representing the user was always visible, conveying a passive message about the user's location.

The clocks fitted themselves to the behaviors of the human body just as did Glancephones, but the fitting we had in mind here was the idea that people who are together can see at a glance who is with them. When they go elsewhere, however, the person who is left can't see where they are. Bricks and mortar and the brutal facts of distance take them beyond view. We thought that Whereabouts clocks would allow the users to break through these walls and the occlusion of distance so as to see their friends wherever they are. Moreover, by linking the location system to messaging, senders of messages could also convey their location without a word. In both regards, we thought that we devised a fitting of the human body to a virtual world—one that allowed one body to see where others were or to let others know where one is without effort, irrespective of the fabric of the real world. The fitting in this case was of the mechanical lookings and glancings of the human body to a representation of other's doings as seen virtually.

This piercing of borders that transcends physical limits and skins has excited many commentators. William Mitchell, for example, has argued that this relationship—the mixed space of the geographically real and the virtual—is the single most appealing aspect of the digital age (see his *Me++* of 2003). Some of the big technology players have also started offering services that make this link (Google's Latitude service comes to mind). Although Mitchell writes about breaking all boundaries, we thought that such virtual lookings and knowings would not have much value if the user could see anyone, since presumably they were interested in only some people—their friends (as in the case of Latitude), say, or their family. We decided to look at family life, since knowing where other family members are seemed to be an obvious a concern. We knew also that knowing the location of others can be a delicate matter, even within families. Teenagers seek to be invisible to their parents while knowing where their parents are, for example, just as the reverse was the case when those teenagers were young and their parents worried that they might get lost. For these reasons, we designed the clock to offer fairly general location information, hoping that in a research study we would learn what level of accuracy might offend and what might be broad enough to be acceptable. We thought too that linking this to messaging might be economical for families, since there would be less need for "Where are you?" communications, the kind undertaken when parents want to know why the kids have not come home or teenagers want to know whether their parents can pick them up.

FINDINGS

We placed Whereabouts clocks in half a dozen households near Cambridge, and we found that knowing the location of family members was more appealing than we had imagined. As we interviewed our research study participants, we learned that the clocks let them gently deepen their knowledge about their families. But although they had better knowledge, what mattered was what they did with this knowledge. This wasn't fitting; what the clocks enabled might be better thought of as greater artfulness, particularly the ability for members of families to be more artful at doing family life.

In one house, the movement of the mother's icon from the Work zone to the Other zone at a certain time of the day (at the end of afternoon) was seen as indicating to those in the house (most often teenage children) that she was on her way home. The teenagers would see this, start boiling water in the kettle, and would have a cup of tea ready for her when she walked in the door. Even if she was late in doing so, perhaps stopping to buy something or getting stuck in a traffic jam (events that often occurred, apparently), the teenagers felt virtuous for having done so. Teenagers might be surly and resentful about some matters, but in this household, these teenagers liked to do this. It was their way of being affectionate. And the mother felt that she was receiving their affection and delighted in it. She remarked on how pleasant these little actions were. Whether tea was actually provided didn't seem to be as important to her as were the teens' intentions. The issue here wasn't communication. It was performance—a doing rather than simply an information exchange. Communicating some information was a prerequisite, but the value of this communication was measured by the act that ensued—in its consequences.

This example of tea making might seem evanescent, but this and other families felt delight when they glanced at the clock and noted where family members were—when, by and large, the location of family members was in no doubt. In most instances, the clock did not provide anything that our users did not already know. My colleagues and I were perplexed as we tried to understand the value (delight, pleasure, and assurance) that these people felt when looking at the clock.

When we first started gathering data from the research study, we found it peculiar that people would look at the clock for evidence about where their siblings and parents where when they usually already knew where these others would be—at school or at work. Occasionally, using the clock to discover where their fellow family members were made sense, but our users seemed to delight in it even when they were sure of where everyone was. Our first reaction to finding this was to ask what value the users were seeing in the clock. We came to learn that the issue was not the digital content in question (the stuff exchanged and made visible) but the reassurance and sense of certainty that this provided. As one user put it, "It makes you think the world is all right, it's all okay." These were doings, too, even though they were less demonstrative than making tea. But our subjects made it clear to us that this aspect of being part of a family had value. The context of family life, we came to realize, was not merely a geographic domain (a particular built locale—a home—that might be fitted to a virtual one). Context here meant how members of the families in question conducted themselves with an orientation to the welfare of their fellow members. If everyone else was in the right place, so to speak, then the moral burden of needing to worry about them was relinquished and dissolved.

In family life, most members of a family do the same thing on most days. Life is mostly routine for families, just as it is for everyone else. But the clock allowed people to know each other in richer ways, and this knowing altered what they did and what they felt obliged to do (they could make tea or worry why people weren't where they should be), and sometimes it let them not do anything (they could forget their family responsibilities for a while). The actions that they took and their perceived character as fond or as thoughtful (as, say, teenagers who either made an effort or cared only for themselves) were in part enabled by the information and sense of the world that the clocks provided.

As we did our research, we came to see that although families might not plan and monitor every movement the way military organizations do, part of family membership involved having a mutual awareness of what other family members were doing. Family members figured out ways to be aware of each other and to show tenderness toward each other. The clocks were one resource that they used to manufacture this sensibility.

LOOKING TO TELL

But what do performance, mischief and laughter, a sensibility for friendship and family life have to do with communications technologies and overload? And how do these facets of human behavior help us to understand what motivates the communication act in ways that can encompass all that we require for our questions?

In sociology, an ongoing debate has centered on the construction of character. For many commentators, human character is a narrative, and human tellings are the vehicle through which identity is constructed. In its simplest form, this interpretation says that humans are creatures who desire to tell their stories. People certainly do delight in telling stories about themselves and enjoy hearing stories told by others. Glancephones and Whereabouts clocks allowed our study participants to construct stories about themselves, their friendships and families. The stories told through the Whereabouts clocks and the Glancephones were bound to the time and place where these channels were used, but within this compass the devices provided different modes of expression. The same story couldn't be told over both devices, both devices couldn't be part of the same story, the same kinds of character traits couldn't be highlighted or performed in both.

One problem with theories of narrative is that they tend to strip out any real sense of the time and space of narrative acts—of where the tellings get told. The bulk of the narrative literature is concerned with the theory of narrative form. The French sociologist Pierre Bourdieu argued in his *Outline of a Theory of Practice* (1972, translated in 1977) that one needs to avoid such distraction to understand humanness. He urged investigations of how there is both a bodily and moral fitting of the human to times and places. Telling stories is as much a question of telling the right story at the right time and place as it is about any general imperative to tell stories. In his view, people have to learn to behave in certain ways in certain places and differently in other places, and telling appropriate stories is one of the skills bound up with this lesson. Bourdieu suggested that the differences in appropriate modes of behavior aren't simply matters of thought or will. He argued that they are also written into the skills of the body and are manifest in movements that fit the body and its movements to the objects that it interacts with. Like the Turing theoretic emphasis on the body (which I took an aversion to in chapter 3), Bourdieu emphasizes the bodily movement of humans and urges us to look at movement rather than at something internal in the mind. But in his *Theory of Practice*, Bourdieu was trying to counter the dichotomizing view of the human that splits mind and body, emphasizes only the body and its actions or movements, and leads to its opposite in anthropology and sociology—a kind of mentalism, a concern with what goes on inside the head without a reference to the body.

As noted in chapter 3, Talcott Parsons (1937) tried in the 1930s and 1940s to counter behaviorism with his voluntaristic theory of action.[4] Irrespective of the debates that this theory has sparked since it was first introduced, it led to an emphasis on how people orient their moral hopes to the material values of society. Parsons's attempts to avert behaviorism nearly removed the breathing body from society altogether.[5] This spurious reductionism is what Bourdieu was trying to combat some forty years later—that it makes no sense to emphasize body over mind, mind over body, or indeed any similar dualism. Bourdieu proposed that this entwining of mind, body, and social-temporal location of performance could be thought of as a *habitus*. This term allows us to avoid confusing words and categories that emphasize either the mind or the body.

Consider how one quickly picks up the phone when it rings at work because the bodily tempo of work is quick, urgent, and responsive. In other spaces, one reacts differently. At home, the household phone might ring for some time, and we might shout, "Who is going to answer the phone?" We usually don't answer quickly or urgently because at home we are oriented to a different set of bodily practices—to being leisurely, indolent, and relaxed. We might not want to be bothered answering the phone. We might hope that someone else who has more energy than we do will pick it up. Sometimes this domestic inertia results in no one answering the phone, and callers leave messages on an answering machine even when they know that we are at home. The phone may sometimes be ignored at work, too, so the two places are not entirely distinct. But the systems of appropriateness and propriety—Bourdieu's embodied action, which merges intention and action—that are appropriate to each domain are different. Each is a different habitus.

Our Glancephone studies uncovered forms of friendship—its mindfulness, bodily arts, and linking of times and places. We saw the habitus of friendship when communications were made between restaurants and workplaces, for example, and when study participants discussed girlfriends, alcohol, labor, and insouciance. Time, place, body conduct, and topicality were elements here. With the Whereabouts clock, our sense of family habitus was sharpened by describing some of the sensibility that family life requires. My colleagues and I tried not to distinguish ideas and actions in the behavior of the research study participants, and we tried not to separate the times and places in which those ideas and actions occurred. Our goal was to grasp what the intertwining of the two, the body in the mind, the mindfulness in the body, were correctly treated as one.

But in so doing, we were also aware that understanding the apparent success of the Glancephones and the Whereabouts clock requires that we understand how a habitus might evolve. The introduction of new technologies will be one of the factors creating this evolution. Consider how it was that in the late 1980s and early 1990s, receiving a mobile phone call in a public place was thought to be rude by many people, and the people who answered those mobile phones sometimes preened with a kind of celebratory vanity. But the codes associated with phones have shifted, as have the arts required for dealing with them. Today users might leave a phone on a table, allow it to ring, and glance at the caller identification number or name on the screen before answering or ignoring it. They use their eyes as much as their minds to make a judgment, and they need to be able to pick up the phone and press either "Accept" or "Call forward" promptly and elegantly. They would be laughed at if their actions led them to spill a glass of wine, just as they would appall their fellow diners if they answered the call by shouting.[6]

In a similar manner, Glancephones and Whereabouts clocks did not resist or transform the social setting in which they were used, nor did they fit a prior need or bodily pattern (even if needs and patterns did help us initially conceive of the devices). They were brought into social settings, and gradually they helped to shift the codes of appropriate bodily *and* mindful behaviors within those settings. With Whereabouts clocks, the habitus of home life evolved as digital icons of location were used as resources that allowed artful bodily doings (for making tea, for example, or for ignoring the world and idly watching television). Similarly, Glancephones did not fit into a cognitive need (related to friendship, say) but were managed so that they gently, skillfully, and playfully expanded and evolved communicative finesse in public and private spaces so that the identity of those involved could be crafted in new light. These technologies gently shifted human doings because humans used them to craft their doings in new ways.

WHY MORE?

Both these technologies were modest affairs, and our research studies of them involved only a handful of participants. But these studies illustrate how we can understand the role played by communications technologies in everyday life. Just as the sociological studies we reported in chapter 4 noted that the role was subtle and diverse, so these show how new modes of communication alter the landscape of human connection. They offer delight, assurance, and opportunities for demonstrations of affection and mockery. We see too how these effects need to be understood in terms of how technology use is bound to physical locale where modes of bodily competence are framed. Motivations behind acts of communication have their proper places, one might say.

These studies can help us get a better grasp of what it might be that entices people to new forms of expression. Their value for users was in helping them distinguish differences in their daily doings—in the doings of friendship, for example, in the affairs of family. The roles that these tools played was varied. No single doing (somehow transformed into communicable materials) could be captured, compressed, and sent via Glancephone and the Whereabouts clocks. Lots of doings were possible with them. The technologies succeeded because users could deploy them in particular ways, with particular goals in mind. In both cases, central to these goals was the portrayal of the one using the technology; the communicator or the user of communicated material. The *who* in the communication act was central to what Glancephones and the Whereabouts clocks enable.

This cannot be said to be the case with all communication technologies. Many communications channels are designed and developed for military organizations, for example, and here the who of the soldier (say) is of no interest. It is what that soldier communicates that is. In these contexts humans act as agents of information, not as agents of their own concerns. At work, our diligent responses to emails are designed to show our professional competence, irrespective of our selves. But in other circumstances (other habitus), we communicate precisely because we want to say something about who we are.

There are evidently metrics here. Users of Glancephones did not convey more about themselves by using the Glancephone more. Their character was conveyed in part by the extent they chose to use it, certainly. Glancing more said something about who they were, and glancing less said something different. It said something about how much they liked to glance, to be glanced at and to laugh. How the glancing said less or more is not a quantitative but a qualitative question. The same could be said of users of the Whereabouts clocks. Acting on an indication that a family member was coming home was something that could be interpreted in terms of frequency—but putting the kettle on every time could be seen as excessive; wasteful even. Doing family tenderness is a difficult task subject to mistakes.

What one learns is that a motivation behind certain communication acts is to convey the adroitness of those involved. It is not merely their adroitness at using the technology in question. It is a measure of their adroitness as people. The studies of Glancephones and the Whereabouts clocks highlight how identity is conveyed in the way in which communication channels are used. To use more channels is not necessarily better, for example, since a way of delineating identity is by assessing how astute that use maybe. Sometimes using a channel a great deal may be appropriate; sometimes the opposite. That this is so will mean that a user will need to reflect on what the use of one channel will achieve and what the neglect of another will avoid. The choice of one channel (or a set of channels over another) is used by others to judge who we are. In this view, identity is bound up with how people choose to express themselves, and in the digital age, this means how people communicate using mobile phones, emails, social networking sites—and new technologies like Glancephones and Whereabouts clocks. In this view, people are *how* they communicate, not something separate from the communication act itself.

Bourdieu offers in his book *Distinction: A Social Critique of the Judgment of Taste* (1984) a reconsideration of habitus and suggests that it evolves over time as people create distinctions between themselves and others. People seek uniqueness in their evolving patterns of action and in their various habitus. Though Bourdieu did not discuss communications channels, this argument applies to understanding why people seek new channels and new modes of communication (even as they worry about taking on too many). They are not thinking about, say, efficiency or economy when they do so—as if these terms were applicable without reference to them, the actor. Instead, they are thinking about what these channels and modes say about them and their worlds. Labels like *economy* and *efficiency* might apply in understanding their efforts at distinction, but these are being used for moral categories and not for quantitative ones. They are only part of the vocabulary that applies to these behaviors. To be economic in communication is to be concise, for example; but concision can easily turn to curtness. The opposite of economy is *prolixity*. In-between these measures are a whole vocabulary of different accents each of which convey something about the one doing the expressing.

CHOICE

People choose new technologies of communication because they allow them to do what they have always done but in new ways (gifting for example), partly to convey a sense of themselves, and partly to become more distinct. Their choice of communications channel is expressive of them. Such choices won't be confined to acts of communication that are ostensively about activities where the delicacies of identity might seem very important (as when a message of love is sent either in a hand crafted letter, in pen and ink, or via SMS, for example), but even in family affairs, where the identities in question are taken for granted. Sometimes the choice of channel might even say something about the group that one wishes to identify with—as a means of celebrating one's own family membership, for example, or as part of a gang of like-minded bloggers. But some modes of communication emphasize the anonymous, the lack of importance of the sender, though identity will never quite disappear from view. These differences are bound to the technology of the channel as well as the motivation of the user; they are also bound to the place in which the channel is used—the habitus. As remarked, work email emphasizes the role of the participants rather more than their individual identity; email used for personal communication quite the opposite. Here *who* really does matter. But where these communications are received matters too. At work, a personal email suggests someone who wants to escape the burdens of labor; at home, a work email one who cannot say no to that same labor. To some degree, the *who* of communication is bundled with the *how* of communication; it is also bundled with the *where* of it. We are not just what we say but also how we say it and where we say it.

In chapter 1, I note that the conventional wisdom says that we seem to have reached a threshold where our scale of communication leaves little room for doing anything else. What one might say about technologies like Glancephones and the Whereabouts clock is that they don't appear to make a great deal of difference to practical affairs but do provide opportunity for the self. With them, we might have reached a point where we lose sight of what is conveyed when we express—other than our finesse as communicators. How we communicate is now becoming a synonym for who we are. And our desire for portraying ourselves adds to the burden of communication.

This is to exaggerate the lessons from studies of Glancephones and the Whereabouts clocks. But the question of motivation is clearly central to any attempt to explain the current communications landscape. Many theories of communication imply things about human nature that seem to be helpful for a discussion of communications technologies, but many of these suffer from a limited view of human communicative practices that don't properly explain this fact—that who we are and how we are seen is, in part, through our choice of communication mode. And our desire to be seen and noticed, to be acknowledged, is also a motivator of being in touch. Thus, if communicative practices are leading us to overload, it is because the various motivations that drive communication are now combining to produce a landscape where those motivations are blurring. We cannot see each other for the sheer volume of messages. Our identity is to be found in the messages, but there are too many to digest. And if there is something sought for in a message, a favor asked or some state of affairs declared, we cannot see this because the message reminds us that there is someone behind it, and that who obscures the what. And beyond that, if we use place as an index of messages, our technologies are allowing us to make that obscure too—we find it hard to distinguish a message sent from work because it is about work, for example, and a message sent between friends that is about friendship. If we have a sensibility for the adroit fitting of actions to contexts, to *habitus*, then our communications rich world is making a muddle of that too.

CONCLUSION

In chapter 4 we saw that the importance of communication acts is so great that sociologists inquire into the consequences of their alteration. Ordinary people also worry about ensuring that communication achieves what they want it to do. Expressing is a serious business, even if it entails laughter. But if it is the case that people take seriously the functions of communication, will it also be the case that they worry about there being too much communication conveyed over too many channels? Just as some communication acts fail, so others give the impression that the person undertaking the act is more interested in portraying themselves than in communicating some thing of value to others? One can hear a contemporary reformulation of the phrase "some people like to hear the sound of their own voices" with the complaint that "some people like to post on their social networking site too much." Likewise, one will not be startled by complaints that some people use new channels "for the sake of being trendy." In either case, the communications in question are viewed as having no value, being merely statements about the creator.

In situations like this one would expect that people would start looking carefully at particular acts of communication to see if they contain anything worthwhile. If some new channels encourage people to use those channels simply to be fashionable then it might be the case that the acts enabled are pointless. Eventually, these new channels will loose their appeal even to those seeking novelty. But presumably the vitality of some new channels is a function of the fact that they do enable acts of communication that are more than displays of self.

Concerns such as these may well lead people to examine the words conveyed in a message, and look less at what it says about the sender. But looking at words can be dangerous. When my colleagues and I first started looking at ways to understand expression (motivated by our trials of Glancephones, for example), linguistics seemed appealing as it seemed to offer a way of grasping words. We imagined that it would be much closer to what we needed than Turing theoretic computer science, for example, or Shannon's communications theory, both of which disregard semantics. But even though this discipline had language at its heart, it seemed to lead us away from what we sought.

Linguistic academics seemed to be aware of this paradox. As Roy Harris, former professor of linguistics at Oxford, explains in *The Language Myth* (1981), many linguists offer a vision that recollects John Locke's *An Essay Concerning Human Understanding* (1690), in which he claimed that communication merely transfers thought from one person to another. This sounds plausible but leads to what Harris calls the *telementation fallacy*, which has a number of key ideas bound up with myths about language. One of these holds that spoken and written words ought to be understood in terms of their technical features and not in terms of their meanings or outcomes; i.e. not in terms of the words themselves.

It became clear to us that the telementation view wouldn't apply in our research. But it also became clear that nor would analogues of it for those people seeking to judge the merits of a communication in the words it contains. Whatever doubts one might have about the merits of some communications channel, people communicate for all sorts of reasons, and only some of them have to do with the exchange of content (words, sounds, intentions, code, and so on). A focus on the words might be motivated by a desire to disregard the sender of a message, but words in a message do many things bound up with what the act of communication in question is part of.

When people talk about the weather, for example, their interactions are not always about exchanging information. Sometimes it is simply to avoid silence. The words really don't matter, but that they fill up a void does. People communicate for fun, and this too can hardly be said to be an act of transference, except insofar as a joke might sometimes be passed from one person to another. People sometimes communicate to tell stories, and although some stories might be said to weigh us down, they are not real objects exchanged between the minds in question. Besides, story telling has many forms, and some stories are designed to be forgotten as soon as they are told. *Giving and taking* might not be suitable here, but *wiling away time* might be.

These are examples of unremarkable activities. Nothing sinister is happening when people communicate about the weather. But communication can involve betraying (or keeping) secrets. Asking some questions might appear to passing the time of day, whereas it is in fact prying. People can disguise their motivations in the words of a communication. They can pretend and they can be dishonest.[7]

The reason that many views about human communication (such as this telementation view) are wrong, Roy Harris explains, is that they don't want to find out what people do when they communicate. They fail to understand what people are doing when they communicate because they get distracted by the stuff that is communicated—words, phrases, language, and hence synonyms for all of these things.

A peculiar consequence of this distraction is that concepts and categories about communication are not only dubious but often the source of further distraction. As Harris notes, examples of this include the distinction between the word *language* as used to define a particular (usually regional) system of words (such as "He studied the origins of the French language") and the word *language* as used to define a general system of human speech (such as "People use spoken language to communicate"). Researchers have investigated the essential properties of language, which has led cognitivists like Noam Chomsky (see, for example, his book *Cartesian Linguistics*, 2002 [first edition, 1967]) to seek surface and deep grammar, for example, and his computational followers to seek transformational grammars of this kind of language. In the work of Ferdinand de Saussure (see his *Course in General Linguistics*, reprinted 2001) and other structuralists, this view of generic language has led to studies of semiotic systems and forms of hidden patterns. It has also led to campaigns to improve regional languages (as French or English), especially for some visions of science. Words of particular languages are lowly technologies (some philosophers, linguists, and psychologists claim). They are used by minds speaking in inadequate and flawed argot and are designed simply to communicate between heads speaking particular tongues. What they seek is a language of science. Both the British philosopher A. J. Ayer (see his *Language, Truth, and Logic*, 1936; for discussion, see Magee 1978) and the American philosopher W. V. O. Quine (*Word and Object*, 1960; for a review, see Dilman 1984) took this approach.

The cognitivists, the linguists and these philosophers all hold variously *distracted views*, and they all focus on not the purpose or expressive value (the human bit) of communications but merely the stuff that is communicated (language, words, code, or whatever). In this view, why we communicate is not as interesting as what content is exchanged. The idea that people might need to send messages to each other as if across a vast divide from inside one head into another does sound appealing, even if it is wrong. It evokes the idea that people might have a cargo that they can share with others and that they might indeed have something worth saying. But there is a difference between the moral valence of a spoken word and the form of its delivery as binary code or electromechanical signal. People communicate for all sorts of reasons, and sometimes their motivations seem ephemeral, but whatever the fact of the case, what we should recognize is how complex the interpretation of an act of communication might be. If some linguists, cognitivists and philosophers chose to ignore the moral valence of communication and end up making errors in their thinking, then how much more likely is it that ordinary people struggle with interpreting the meaning of a message? That all people have an ability to make judgments about who we are, what we want to do and where we are through the prism of an act of communication doesn't mean that they make the right judgments all the time. How we express things in a spoken word, a displayed signal, or a received gift are all matters to do with being human, but it is human also to recognize that mistakes can be made as regards understanding what those acts "mean." It is human to make mistakes. As we confront a world were more and more messages and communications are exchanged, so the likelihood that mistakes will become consequential increases. Communication overload might be a label that implies arithmetic, but it turns out to label much more.

NOTES

1. Discussions of identity and digital networks, perhaps encouraged by the success of books like Sherry Turkle's *The Second Self* (1994; reprinted 2005), tend to take an exaggerated line. For a more judicious view, see C. Greiffenhagen and R. Watson's 2005 article "Theory and Method in CMC: Identity, Gender, and Turn-taking: An Ethnomethodological and Conversation Analytic Approach."

2. In the mid-1990s, this pairing was thought to be a suitable basis for the design of interaction dialogs with computer kiosks. See Wooffitt et al. 1997. Wooffitt and colleagues' book led to a ferocious debate about whether such systems were mechanical or normative. If mechanical, then a machine could be programmed in a similar manner; if normative, then it could not.

3. This is actually something I had been arguing for some time before we actually got about inventing something to address this issue. See my own 2003 paper, "People versus Information: The Evolution of Mobile Technology."

4. Parsons became increasingly interested in systems theory and cybernetics. His later views resonate with the kind of simple rendering of human nature that I describe as the basis of inventive reason in chapter 3. For a review of Parson's intellectual development, see Collins and Collins (1973).

5. The opposite tendency in sociology took the behavioral side. As G. H. Mead (1964) noted, symbolic interactionism could be seen as a kind of social behaviorism.

6. See especially Hamill and Lasen (2005) and Harper, Palen, and Taylor (2005).

7. This is a fundamental aspect of social organization, as many sociologists have pointed out. In sociology, a related distinction is between gemeinschaft and *pseudo*-gemeinschaft—between a real commitment to shared community and a false one (see Merton 1957). Research studies rely on participants who don't hide their intentions, but at

times people do hide them (for all sorts of reasons). The basic distinction applies also to arguments about the function and moral order of communications (which we return to in the final chapter) that are central to the argument put forward by John Durham Peters in his 1999 *Speaking into the Air: A History of the Idea of Communication*, where the desire to see only the real person is a motivator behind the adoption of various new communications media. Seeing someone via video may persuade you to trust them more than if they sent you an email, for example.

REFERENCES

Austin, J. L. 1962. *How to Do Things with Words*. Oxford: Oxford University Press.

Ayer, A. J. 1936. *Language, Truth, and Logic*. London: Penguin (reprinted in 2001).

Bolter, J. D. 1991. *Writing Space: The Computer, Hypertext and the History of Writing*. Hillsdale: Erlbaum.

Bourdieu, P. 1972. *Outline of a Theory of Practice*. London: Cambridge University Press.

Bourdieu, P. 1984. *Distinction: A Social Critique of the Judgment of Taste. Trans. R. Price*. Cambridge: Harvard University Press.

Brown, B., N. Green, and R. Harper, eds. 2001. *Wireless World: Interdisciplinary Perspectives on the Mobile Age*. Heidelberg: Springer Verlag.

Chomsky, N. 2002. *Cartesian Linguistics: A Chapter in the History of Rationalist Thought*. 2nd ed. Cambridge: Cambridge University Press.

Collins, O., and J. Collins, eds. 1973. *Interaction and Social Structure*. The Hague: Mouton.

Conlon, T. 1985. *Start Problem-solving with Prolog*. Reading: Addison Wesley.

Derrida, J. 1974. *Of Grammatology*. Trans. G. Spivak. Baltimore: Johns Hopkins University Press.

Derrida, J. 1977. *Limited Inc*. Evanston, IL: Northwestern University Press.

Derrida, J. 1988. Signature event context. In G. Graff, ed., *Limited Inc*. Evanston: Northwestern University Press.

Diamond, C. 2008. *Philosophy and Animal Life*. New York: Columbia University Press.

Dilman, I. 1984. *Quine on Ontology, Necessity, and Experience*. Albany: State University of New York.

Goodwin, C. 1979. The interactive construction of a sentence in natural conversation. In G. Psathas, ed., *Everyday Language: Studies in Ethnomethodology* (97–122). New York: Irvington.

Greiffenhagen, C., and R. Watson. 2005. Teoria e método na CMC: identidade, género e tomada-deturno: uma abordagem etnometodológica e analítico conversacional (Theory and method in CMC: Identity, gender, and turn-taking. An ethnomethodological and conversation analytic approach. In A. Braga, ed., *CMC, Identidades e Género: Teoria e Método* (89–114). Covilhã: Universidade da Beira Interior.

Hacker, P. M. S. 2007. *Human Nature: The Categorical Framework*. Oxford: Blackwell.

Hamill, L., and A. Lasen, eds. 2005. *Mobile World: Past, Present and Future*. Godalming, UK: Springer.

Harper, R. 2003. People versus information: The evolution of mobile technology. In L. Chittaro, ed., *Human Computer Interaction with Mobile Devices* (1–15). Berlin: Springer.

Harper, R., L. Palen, and A. Taylor, eds. 2005. *The Inside Text: Social Perspectives on SMS*. Dordrecht: Kluwer.

Harris, R. 1981. *The Language Myth*. London: Duckworth.

Hiltzik, M. A. 1999. *Dealers of Light: Xerox PARC and the Dawn of the Computer Age*. New York: HarperCollins.

Jefferson, G., and E. Schegloff, eds. 1992. *Sacks: Lectures on Conversations*. Oxford: Blackwell.

Jones, S., ed. 1998. *Cybersociety 2.0: Revisiting Computer-Mediated Communications and Community*. Thousand Oaks, CA: Sage.

Landow, G. P. 1992. *Hypertext: The Convergence of Contemporary Literary Theory and Technology*. Baltimore: Johns Hopkins University Press.

Locke, J. 1690. *An Essay Concerning Human Understanding*. London: Penguin (reprinted 1997).

Magee, B. 1978. *Men of Ideas*. London: Oxford University Press.

Mead, G. H. 1964. *Selected Writings*. Ed. A. Reck. Chicago: University of Chicago.

Merton, R. K. 1957. *Social Theory and Social Structure*. Glencoe: Free Press.

Mitchell, W. J. 2003. *Me++: The Cyborg Self and the Networked City*. Cambridge: MIT Press.

Parsons, T. 1937. *The Structure of Social Action*. New York: McGraw Hill.

Peters, J. D. 1999. *Speaking It the Air: A History of the Idea of Communication*. Chicago: University of Chicago.

Quine, W. V. O. 1953. *Word and Object*. Boston: MIT Press.

Saussure, F. de 2001. *Course in General Linguistics: An Anthology*. Ed. M. Ryan and J. Rivkin. Malden, MA: Blackwell.

Shannon, C. E. 2001. A mathematical theory of communication. *ACMSIG Mobile: Mobile Computing and Communications Review* 5 (1):3–55.

Shirky, C. 2008. *Here Comes Everybody: The Power of Organizing without Organizations*. London: Allen Lane/Penguin.

Sudnow, D. 1972. *Temporal Parameters of Interpersonal Observation*. In D. Sudnow, ed., *Studies in Social Interaction* (259–279). New York: Free Press.

Turkle, Sherry. 1984. *The Second Self: Computers and the Human Spirit*. New York: Simon and Schuster.

Turkle, Sherry. 2005. *The Second Self: Computers and the Human Spirit, Twentieth Anniversary Edition*. Cambridge, MA: MIT Press.

Wiener, N. 1948. *Cybernetics: or the Control and Communication in the Animal and the Machine*. Cambridge, MA: MIT Press.

Wooffitt, R., N. Fraser, N. Gilbert, and S. McGlashan. 1997. *Humans, Computers and Wizards: Analysing Human-Simulated Computer Interaction*. London: Routledge.

6 THE SEAMS THAT BIND

PREAMBLE

Although social science and philosophy offer some conceptions that lead us away from an understanding of human communication, sociology and linguistics offer approaches that can allow us to build on our studies of Glancephones and Whereabouts clocks. We can see that the moral values of communication acts are important, just as are the places where those acts take place. The social and moral location of an act of communication—its coordinates in space and time—can be thought of as a habitus. And we can see that individual distinction is managed by choosing to communicate in one way rather than another. Lastly, people communicate sometimes because they have something to say and sometimes simply because they desire to express themselves for expression's sake.

These meanderings seem to lead us back to philosophy—not to the ideas of Locke, Quine, or Ayer but to a philosophy that asks questions about what is good for us, our society, and our aspirations. Much philosophy appears to be about precisely this subject—the study of the goodness (or otherwise) of our ambitions. Friedrich Nietzsche, in *Thus Spoke Zarathustra* (1961) and elsewhere claimed that philosophy ought to provide a moral compass for the secular soul. But moral philosophy is only part of what philosophy entails. Another, perhaps larger part involves what philosophers from the time of Plato have argued comes before morality—the conceptual antecedents of ideas and their categories. Philosophers are interested in understanding how we come to have ideas about things like morality and categories like good and bad *in the first place*. When we have a precise sense of how the concepts operate, then we can use them to work for us. Roy Harris (1981) and Pierre Bourdieu (1972, 1984) offered my colleagues and me guidance in our research into the Glancephones and Whereabouts clocks, but our research showed that we didn't need science or academic psychology to understand the significance of a cup of tea or to appreciate the fun of being glanced at. We needed to use the expertise about the world that we had gained by living *in* that world. This expertise let us see the difference between laughter and seriousness or between affection and indifference in the ways that Glancephones and Whereabouts clocks were used. This expertise can apply to any and all technologies that let us *keep in touch*.

My colleagues and I gradually came to this point of view by moving away from our initial assumptions about what communication was and how it ought to be understood (mentioned at the outset of the book). This radical move meant that we had to abandon our inventive approach to the communication act and scientific approaches that sought to reduce communications into something other than moral acts. We eventually settled on a commonsense or real-world approach to communication that allowed us to understand (and hence analyze) an act by what it implied about the actions of those involved and hence its moral consequences.

Some people might feel ill at ease when presented with a choice between what appears to be science (rational inquiries) and common sense (folk wisdom). Our argument, after all, has led us to turn away from one of the great achievements of the past two hundred years—science. But the way that people communicate to each other cannot be understood by science because intentions and meanings are the topic and science is not good at these phenomena. Because expressing is a performative act with moral implications affecting the intentions and subsequent acts of those involved, the right approach to the analysis of such acts must place that moral element at the center of attention. For this reason, the Royal Society's credo, *Nullius in verba* (nothing on another's word), ought not to apply to the inquiries we are undertaking here. Our concern is with what people say, why they say it, and what the consequences of their saying it are. Ours is more like a philosophical anthropology, as the Oxford philosopher P. M. S. Hacker proposes in his book *Human Nature: The Categorical Framework* (2007).

Because we use our knowledge of the human world to explore that same world doesn't mean that it is easy or that we cannot make mistakes or get misled. We might need to watch for the traps that words can induce (as Harris warned), and we might need to reassert the significance of place and embodiment in communication despite the temptation to neglect these issues (as Bourdieu noted). Beyond this, we need to distinguish between questions that are appropriate to ask about communication and those that would be odd and not helpful to ask. When people express something, they don't ask what expression is, for example, but assume that everyone knows what an expression is. When people listen to someone else, they don't wonder what listening entails. They already know, and it is only in that way that one can listen. Certain parts of communicating are so basic that we don't really doubt them even as we perform a communication. The gist of the prior chapters, particularly the last one, is that communication (as conceived of for our inquires) is not a physiological topic but a conceptual one having to do with what listening means, what expressiveness presumes, and what communication enables.[1]

Thus, we have returned to philosophy—looking at the assumptions that underscore particular kinds of act. In this chapter, I am also concerned with assumptions, the ones underscoring acts of communication. The reader might wonder why I want to address this topic now, in the penultimate chapter. Have we not learned enough from our previous inquiries to approach the topic at hand? Thus far in this book, we have been able to refine *our own sensibilities* for the humanness in communication. Just as my colleagues and I had to work at reconfiguring our inventive imaginations to understand the unexpected uses that our technologies were subject to, so we have had to work to understand what we already know, as ordinary people who communicate a great deal, if we are to start making judgments about our own communications and those of others. The things that apply to judging our communication acts are ones that we don't ordinary deal with. We take them for granted, even as we busy ourselves expressing.

This can perhaps be better conveyed by returning to the metaphor that is the title of this book. Communicative practices create a texture—a complex weave of bonds that tie together those who are communicating. This texture has various forms and strengths: some bonds created through acts of communication are instant and others slow, some ephemeral and others more permanent. These bonds vary according to the type of act in question and in terms of the technologies that are used to enable acts (although technological mediation is not a prerequisite of acts so much as a property of some). In general, the view I have been trying to develop holds that the bond created by an instant message is different than the bond created by a video call, just as both these are different from a written letter and different yet again from a whisper.

In this chapter, I try to deepen our sense of this texture—not by creating a taxonomy of the acts but by looking at the elemental features of the fabric that they create. Here the allusion to the philosophical task of inquiring into the origins of concepts begins to make sense—looking at properties of communication that allow us to build on these acts and distinguishing between what different acts enable at a level above and beyond those elementals. If we create a texture of relations through our communication acts, then elements of that texture are taken for granted and also are the structural basis of those acts. They are the seams that tie the texture together, the hems that have to be hidden to do their work, and the lining that gives the texture its shape. When we look at a man wearing an elegant suit, we appreciate the cloth's color and pinstripe pattern but don't see the lining that gives it shape and the padding that gives him a masculine form. Like the seams and the hems, we take for granted that these hold it all together and let us focus on him, the man.

A new approach in the human-computer interaction (HCI) literature has suggested that computer systems might be designed to make the seams and hems that combine and link various systems more visible. This would allow users to fabricate and combine the systems as they see fit (see, for example, Paul Dourish's 2001 book *Where the Action Is: The Foundations of Embodied Interaction*). Although I am sympathetic to this view, here I am proposing the obverse—that people strategically choose and interpret diverse acts of communication by taking for granted some aspects of the acts in question. The seams have to be invisible for them to focus on this strategic behavior. But I want to investigate these seams so that we better understand that strategy in the next chapter.

A STARTING POINT: WHAT IS THERE TO TALK ABOUT?

Not much philosophy has been done with communication or expression, even though philosophy often seems to concern itself with language and words. Indeed, as Simon Blackburn notes in *Spreading the Word: Groundings in the Philosophy of Language* (1984), language seems to be the essential topic of philosophy. Communication and expression meanwhile are alluded to obliquely when matters apparently more pressing on the philosophical imagination are considered. Some of these topics do seem closely allied to our own (such as when Plato discusses the differences between the spoken and written word in *Phaedrus*). But the question of *why* something is said is not often considered. Occasionally, it is addressed badly, as with John Locke's (1690) telementation thesis. Communication does come up a great deal (though indirectly) when, for example, the fact that something is said is used as a vehicle to inquire into other matters. John L. Austin uses the words in phrases that might be spoken to explore the relation between meaning and action. Indeed, in books like *How to Do Things with Words* (1962), Austin is concerned with issues similar to our own. Paul Grice's *Studies in the Way of Words* (1989) seeks to look at the assumptions that underscore conversation. A third, John Searle's *Speech Acts* (1969), might seem closest to our interests, given Searle's interest in the performativity of communication and his interest in the work of Bourdieu and his concept of habitus. But Searle is more concerned with developing a theory of meaning rather than with exploring why people communicate. Searle never asks why people seek to keep in touch. He is interested not in the morality of expression but in its facticity—that it is done.

Nevertheless, I start my discussion in the way that most philosophers typically do. Philosophers like to ask what concepts like truth have as foundational properties. I am interested in communication and its foundational properties (keeping in touch). One way to begin is to look at how philosophy looks at the *why* of expression, even if Austin, Grice, and Searle don't do so in the way we desire. Locke offered a theory about what was communicated and where it was sent (from one head to another) and offered an answer to the question of why with his telementation thesis. This was an unsatisfactory view. More modern philosophers have offered answers to the question of why, but their answers are insufficient for us. Nevertheless, looking at what they have attempted to do might allow us to answer our questions more easily.

The philosophers I am thinking of have been asking the question of why as part of a separate discussion of *skepticism*—the idea that the world and everything and every person within it might not be real. It might be a mirage. The skeptic's problem is to prove that the world is either false or real. Skeptic philosophers argue that one way to answer their doubts might be to test whether *others* exist, and they do so by sending out a *communication* (consisting of a description of the world) to those others to see what happens—to find out if their communicated words get a response. If a response is received, the skeptic hopes that it is also a description of the world and corresponds with the one that he or she sent initially. If so, then the skeptic can be sure the world he or she inhabits is the one that is populated by others.

This type reasoning, often labeled *Humean* in honor of the great Scottish empiricist, David Hume, sounds plausible, but in exploring its flaws we can get to our topic. One of these flaws is that skepticism is implausible on simple conceptual grounds. As Stanley Cavell notes in *A Pitch of Philosophy: Autobiographical Exercises* (1994), it suffers from the infinite regress problem. There is no way that the individuals involved in a communication can ever know for certain whether the world that each is talking about is the same as the other's. Even though both might use the same words to describe their respective worlds, each person cannot be sure whether those words describe the same thing. They might, but they might not. What is pink to one person might be yellow to another, and things might be made worse by the fact that both use the word *blue* to describe the different thing in question. They can try other words and other descriptions, but in each and every case, however far they *regress*, there will be no end point—no place at which both parties can be sure that each is talking about the same thing. In Cavell's view, the skeptic's fate is to be left unsure, unable to vouchsafe that any proof of another's existence and the world in general is real proof and not its chimera. If one starts off as a skeptic, one remains a skeptic.

But if people are not skeptics seeking empirical proofs of the existence of the world, then what are they doing when they communicate? Cora Diamond, in her essay "The Difficulty of Reality and the Difficulty of Philosophy" (2008), offers a corrective to what she sees as the oddness of the skeptic's view. In so doing, she begins to offer an answer to our question about why people communicate.

She argues that people communicate for all sorts of reasons but not because they seek an affirmation that they are in a world in common. They communicate from the assumption that they are in a world that they already know they share. This lets people speak about particular things. For Diamond, skepticism is merely something that philosophers have conjured up by imagining a world to be other than it is. It is an exercise in imagination, not a description of how things are or the basis for explaining the auspices behind the expressions between persons. In her view, people express themselves to others because they know they share the world; only philosophers think otherwise.

Although Diamond's target is philosophical skepticism, her argument gives us a clue about how we ought to treat human communication and its requisites. If Diamond is right that the world-known-in-common is a prerequisite of communication, then whatever the motivation of a communication act, it cannot be an attempt at a scientific description (what the skeptic does when he or she sends out a message and hopes for a response). This point will help us avoid certain errors that might affect our efforts to judge real acts of communication (not philosophers' hypothetical acts).

If people aren't in the business of offering sciencelike descriptions, then we shouldn't triangulate their expressions (and whatever descriptive elements they contain) with forms of description that are not communication acts. One order of description is an act of communication and a vehicle for intentions, and the other is a form of inquiry and not an expressive act. Consider how we ordinarily respond to recollections of the past. When we engage in communication acts with people who are recollecting, we listen to them, hear what they say, and do not treat those recollections as if they are faltering attempts at truth that might be supplemented by means that are orthogonal to that act of communication in the first place. We don't say, "Hold on. Let me test your recollection with statistics." We don't judge a person's *recollections* by ideas about randomness and proof and their mathematical properties. We judge them by what we think the person is trying to do in raising or evoking those recollections. We refer to the value that the recollections provide in a certain context—a conversation undertaken in a particular place with particular goals in mind. People's recollections of the past should not be treated as scientific descriptions. They maybe resources for a science of the past, but they are not part of that science, nor are they attempts at doing science.[2]

People sometimes talk about the world in empirical terms. But if one accepts Diamond's assertion, then we should recognize that those engaging in such acts are doing particular work. They are describing some part of the world that is unusual or has special merit, for example. They aren't describing the world because it is strange, as a skeptic would, but they are describing the world because some of its particulars are strange. Similarly, people sometimes talk about the world when they want to explain that it is has changed. But they mean that bits within the world have changed (such as new technologies), not the world as a whole.

Even if one accepts that people don't offer descriptions of these changes as scientific accounts, we do judge what others see and seek to determine whether their accounts are true or accurate. But the ways we do so are bound to the communication acts in question—where they were said, what was intended, and what was claimed. After all, a story might include an empirical description of the world, but listeners don't complain about the adequacy of that description if the story is entertaining. Descriptions offered in communication acts don't have scientific status, but they do have another possible status: they can be artful and creative in their own measure. This too is a foundational property of the communication act: such acts can be art forms. A description can be judged for its accuracy and for its eloquence and charm. These are not the same qualities, and they lead to slightly different sets of criteria for their evaluation. A charming description will be seen as less than a bland empirical one if it is misleading, for example.

One does need to be alert to certain assumptions about communication acts before beginning the business of evaluation, in other words. When one evaluates communication acts, one can ask whether an aspect of the world is as different as someone claims, whether a technology that someone says is new is new for another, whether the changes that someone thinks it is bringing about are consequential for a third, and whether an artful description is artful or talentless. One can ask all these things because we take the world for granted. It is the starting point of our conversations, not the outcome.

WORDS AS PART OF THE WORLD

This is one set of related assumptions that applies when we think of communication acts. But they still don't tell us all the reasons that people talk. If Diamond's view is right and we speak when we see that the world has changed in some detail, then people would talk only when there is something to talk about—some news about changes in the world. But although sometimes one prefers a conversation when news is shared, on other occasions, one simply converses. Conversation doesn't always need to have a point. But this doesn't mean that there is an absence in a conversation that ought to be corrected. Something else about the nature of the world is the starting point—the basis of our knowledge and understanding of what communication acts are.

Communications, talk, and expression (mediated or otherwise) are intrinsic to many activities. The world is not one thing and human communication another. They are often inextricably intertwined. Acts of communication help make friendships, for example. Friendship is not merely a physiological thing, and communications between friends are not simply an incidental excrescence.[3] Families are not merely legal entities. They are doings, too, that often entail what expression allows. This we saw in small measure with the Whereabouts clock. In Ludwig Wittgenstein's description of the bricklayers in *Philosophical Investigations* (1953), he explains that noises, grunts, and sounds are used by the men to indicate what bricks need to be handed over and laid next. In this vignette, he was noting that communication acts are intrinsic to some of the things that humans engage in. It is, as Wittgenstein put it, part of the *forms of life* in question. One might say that talking is part of what we do, not separate from what we do.

This is something that has been written about extensively in the philosophy of language. But a sociologist's thinking on this issue highlights some further assumptions about communication acts that are relevant to our inquiries. Harold Garfinkel, in his

Studies in Ethnomethodology (1967), transformed sociology in the 1960s in large part by pointing out that ordinary talk was part of and intrinsic to the way that the human world is organized. Prior to Garfinkel, sociology had more or less ignored what people express as part of their activities and concerned itself instead with topics that could be examined without reference to the ways in which people spoke, described, or accounted for their actions. Sociology was interested in people's opinions and in what they said—but only as indices of something else, such as their attitude. Garfinkel made it clear that this was missing a fundamental part of human action—namely, how words, descriptions, and accounts are inextricably part of actions. Words can at times describe and organize the very things they are part of, Garfinkel explained. Subsequent to Garfinkel, talk has been much more important in sociology, and this was especially through the seminal work of Harvey Sacks in the 1960s and early 1970s (see his *Lectures on Conversation* published in 1990).[4]

If Wittgenstein and Garfinkel are right, then communication acts are a prosaic feature of the daily doings of people. One should not be surprised that what people say is part of what they do. Indeed, in many human affairs, communication is properly viewed as a normal, commonplace, natural feature of them.

I say this now even though I argued in prior chapters that my colleagues and I have often not treated human communication in this way. We have taken out the prosaic—the obvious facts that make up the why of communication—and have sought to design for a vision of the communicating human that emphasized other properties. For this vision, communication was about fitting bodies together, for example, or expressive richness was attained by sensual richness in the channel—where we sought touch as well as sound and sight. This different vision of humanness had its benefits, offering us pragmatic routes to new invention, but it often led us away from seeing the ordinariness of human expression—the why behind the Glancephones and the Whereabouts clocks, for example, the playfulness of those who used shared whiteboards, and so on. We knew that humans communicate, but we did not allow ourselves to include *all* the reasons for that communication. We could not grasp the full extent of how people communicate *to be* human in the sense that the phrase *being human* ordinarily has—an ability to do the things that people do (work, play, rest, love, lament, laugh, suffer) and to express themselves in those acts. Our vision inhibited our ability to see that people talk to work, express to play, chat to rest, and so on.

This prosaic fact of life makes an assumption about human communication. One needs to understand how communication is part of human action and is constitutive of the society that people make through their expressive acts. How this functions might be prosaic, but it manifests itself in a whole range of human actions that make society rich and diverse. And we need to be alert to this richness if we want to make judgments about what communication is doing, whether or not it is mediated via technology. Indeed, this is especially important when we look at mediated communication, since many commentators lose sight of this. Social networking sites are commonly thought to lead people to meet new friends, expanding the number and geographic spread of their social contacts. As it happens, not many new friends are made via social networking sites, but some commentators on social networking write as if social networking sites are *the only way* that twenty-first-century souls connect with others.[5] Friendships are made in lots of ways and not just via Facebook or dating sites, and the diversity of these ways is a remarkable feature of society. Invariably, these ways entail communication acts—all sorts of acts and mechanisms for creating ties between people.

THE MORAL REQUIREMENT TO EXPRESS

If the world as known in common is the starting point of conversations and if people organize themselves and the world they fabricate through talk—through acts of communication—then we need to think carefully about this diverse and complex world and its many different modalities of expressive performance. Consider how tricky, in terms of expression, some activities are. Some require people to express themselves not to organize the activities but to share an opinion, even if they do not actually hold the opinion. Social bonds are sometimes made through dissembling, and hence society, if it is made through expression, can have peculiar foundations.

The investigations I report in prior chapters have been limited in various ways. In chapter 5, I use the making of tea as an instance of family grace to support my argument that technologies can act as vehicles for being in touch. In chapter 4, I consider whether social relations are changing radically because of the emergence of new communications channels. Many sociologists are convinced that radical change is underway. In their view, the grand sweep of history is loosening people from traditional geographic bonds and lobbing them into spaceless, virtual relations. Such claims are often exaggerated, and the relationships between communications technologies and forms of social relations take are more complex than many commentators would allow. The texture of society, created through communication, is as binding now as it always was, but it has other ways of binding people. These differences (small details about the ways people connect, for instance) are often neglected by sociologists.

The word *prosaic* can encapsulate certain human doings, but it would be misleading for certain types of doing that cannot be undertaken without words—and not because words help organize them. The words in question demonstrate a humanness that is fundamental to those activities. An absence of words in the contexts in question demonstrates an opposite—sterility, coldness, in-humanness—that is hugely consequential. In this respect, these occasions are the opposite of prosaic. These occasions are important not because they are important socially but because they illustrate the amazing variety of communication acts and the ways they can bind people together in surprising ways.

There can sometimes be a particular relationship between certain kinds of event and people's expressions. Sometimes these relations are obvious and commonplace. A person has to say "I do" in a marriage ceremony for the ceremony to succeed, for example. But other occasions elicit comments and expressions—communication acts of various kinds—that tell us that the relationships (between place and expression, meaning and expression, and intention and expression) are complex, linked to the judgment of the persons involved, and thus linked to the bonds that those persons can affect or be affected by.

Some good examples can be taken from expressions and acts of communication that related to love in its many forms and gradations. Consider how words spoken by a friend can sooth the aching of another's heart when that heart has been broken, for example. The words offer encouragement and sympathy. The words suggest not simply that the speaker can understand the aching of the other's broken heart but that he or she understand it all too well. In part, this alludes to communication sometimes asserting the experience of being in the world in common—in this case, a world where the anguish of the human heart is known in common. But it also points to judgments about the bonds between the people involved in the acts of communication. Asserting through an act of communication that one understands another's aching heart is not an empirical claim about one's feelings. It doesn't matter whether one's heart aches or not. The important point is that we say it does. The saying, not the feeling purportedly described, is important. The expression is not about the one who is expressing sympathy. It's about a relationship between two people that requires one to offer sympathy to the other. Friendship is at issue in the act of expression.

Knowing what a communications act implies about social bonds or about those involved isn't always easy, however. Sometimes words of sympathy aren't what they appear to be. In some instances, they are not genuine. Sometimes one is angry with the broken-hearted person since one knew that he or she was not in a suitable match. Often our sympathy is leavened by a simple dislike of the one who has broken the heart of the person we are offering sympathy to. In such instances, our sympathy is really a disguise for fury. In our communication acts, we sometimes disguise our feelings to save someone else's and to honor the bond that makes us friends. Judging or interpreting communication acts is not simple, even though judging them is a natural feature of the society that is made up by those acts. This society is fabricated by nuanced distinctions that lead us sometimes to avoid the truth and sometimes to state it, sometimes to avoid issues and sometimes to refer to them bluntly.

Continuing on the theme of love, someone at a wedding would be viewed as rude if they spoke no encouragement to those tying the knot. Saying something here is more than an obligation or a mechanical act. It affirms a person's understanding of the profundity of the event. Marriage is not taken lightly by those who do it and is not ended easily. When a wedding guest congratulates the newly married couple, these acts of communication acknowledge the seriousness of the event and their recognition that their friends take it seriously too. If they failed to speak, those getting married would interpret the lack of communication to indicate a chilly aloofness and a callous indifference toward the event and their own serious intent.

Not everyone feels the same about marriage, even if all might accept that for some people it is a serious act. Participating in the event might involve acknowledging that the event is serious, but it might also entail hiding one's disdain. So expression is not merely descriptive ("What a nice wedding" or "How beautiful you look") but is a measure of the esteem that you have for another person. And this also holds true in the reverse: your behavior is used to measure or influence how others esteem you (such as "He doesn't care for me, so I won't invite him to our child's christening" or "He was so gracious at the wedding, it would be lovely to invite him to the christening").

The bonds that communications *make* are also sometimes the *basis* of communications, and communication acts can presume the relationship between the parties involved. The relations between persons characterize what is said or done and how the acts are interpreted. Because of this, when we express ourselves, the expressions we choose are a reflection of those ties. We express to affirm those ties, and we express to give renewed strength to old ones. And our acts can also become the reason to end ties.

As the examples of love and its management in friendship indicate, these ties can be delicate and easily altered by communicative performance. They can also lead to negative judgments about that performance. There are times and places when expression is required, and in the same places and times (habitus), saying nothing sometimes speaks volumes too. And all of this can turn on matters that seem beyond reason and almost physiological—heartfelt loves, pain or anger, passionate dislikes for another. Fibbing can bring people together, just as it can throw them apart. But fibbing is done when one knows who one is fibbing to: the relationship one has with the person determines what is a little white lie and what is dishonorable dishonesty.

SOCIALITY

The form of the relationship between the persons involved is crucial to how the act of communication is acted on and understood. In chapter 5, we explored communication acts between persons who already knew each other—between work colleagues with Glancephones, for example, and between family members with Whereabouts clocks. But a particularly salient distinction that wasn't covered is between those who already have a relationship and those who have never met before some communication. Nor did we consider the related distinction between instances when an opening communication leads to a relationship in the future and instances when an opening communication does not lead to future turns. It may be that some communications are not intended to be openings. Nor did we discuss relationships that are asymmetric, where a group listens to the communications of a celebrity or a leader. All of these bonds have consequences for the acts of communication that sustain and invigorate them, even for acts that violate them.

These social bonds are diverse and complex, although these words seem weak when used to describe them. We have focused on the problem of frankness and social bond where there is a presumption of friendship. There are too many kinds of relationship between persons to describe them all, so focusing on the distinction between those who know each other and those who don't will help us uncover some more assumptions of communication.

One might start thinking about this distinction by asking why it would matter whether there is a relationship. Those who have met in the past and created a relationship of some kind have immersed themselves in a moral code whereby they can always renew a communication from the starting point that they have rights to communicate. If one person sends out a communication, even if it is after twenty years, a failure to respond to a communication by the other (assuming that the message finds them) is viewed as communicating something—a judgment about the intentions of the respective parties. It might suggest that one no longer values the other, for example. As was shown with the greeting sequences discussed in chapter 5, silence can be a communication act that is just as powerful as (and sometimes even more powerful than) spoken words or sent messages.

The form of a relationship comes to be a fundamental starting point for interpreting communication beyond the greetings sequences and in the actual body of communication acts. Turns at communication have a rhythm that bestows meaning, but the rhythm here is not in terms of the seconds that elapse when people talk. The rhythms I am thinking of can last for years. They are the prosody of a human connection, not the prosody of a particular mode of expression. Once one has made a shift away from a relationship that is temporary and to one that has duration through time and hence is a different kind of connection, that connection comes to be one that can never dissolve (unless intentionally). One might not communicate for weeks or years, but when communication happens, one is simply taking the next turn in that relationship. The absence of communication in the intervening period might be a topic, but once in a relationship, whenever one of those involved makes a communication with the other and however they do it (by email or by letter, by a social networking site or a phone call), that communication is framed. This frame means that a response is required since a relationship is presumed.

By the same token, communication between strangers is bound up with reference to this system. When strangers communicate to each other, they lack rights regarding future communication, and their demands and adroitness in communications are *in the present.* They are engaging in the job at hand, even if they are making small talk while they are doing it, and are not seeking to open up a conversation that will lead to future turns at talk. They must not treat the job at hand as a pretext for creating friendship, for example.

The differences are great between simple, functional conversations between strangers, conversations that preserve the status of those involved, and conversations that change that status. Graduation from one to another is a socially significant act, even if it sometimes is a source of confusion. Sometimes one party thinks that a first step at friendship has been made, but the other does not, for example, and sometimes people slip from small talk to friendship without meaning to.

When a shift between the functional and the social occurs, something in the perceived intentions of the actors involved changes too. Now there is a view that the relationship has a value that needs to be honored, worked at, and invested in through future turns at communication. Whether those investments turn out to be great or small are themselves used to measure the intentions of those involved. It is not simply turns at communication that are implied, since all turns are oriented to within a larger compass where the requisites of appropriateness are also brought in to play—and these requisites (or prerequisites) are broad indeed. Two people may treat each other as friends, but there are gradations within friendship that affect the order and character of the communications between them. Different modes of behavior are manifest in topics and mutual concerns that help constitute different types of friendship. Some topics are treated as the private concern of individuals and can be breached in conversations only when the person in question chooses to bring up the topics (except in intimate friendships). Other requisites of friendship and sociality are governed by ideas about age, such that friendship between an older and a younger person is viewed and oriented to as different than friendship between persons of the same age. Gender is a constraint on the compass of interaction that is bound up with potential confusions about whether friendship is the first step in a path to romance or merely a friendship. The relative status of these two modes of relationship is a cause of confusion too. The texture of sociality is rich indeed.

WHEN WE SAY TOO MUCH

There is one last set of assumptions about communication that I want to demarcate. When there is no shared understanding of what is right and wrong, judgments about acts become difficult. But when there is, the judgments lead one to think of not arithmetic but the intentions of those who are overloading us.

Consider the complaints about email overload with which I started the book. When email was first introduced, there was much discussion about how effective email was and how it made organizations better able to share knowledge. It would allow the organization as a whole to become more *informated*, to use Zuboff's phrase (1988). But although email is still useful—even vital—to organizational life, there is too much of it.

Such complaints are not related merely to the idea that one communicant has used too many emails to get the job done. Complaints about overload don't usually suggest that the one who has sent too much has simply selected the wrong register (with one register requiring more emails than another, say). Instead, our complaints point toward the idea that too many emails have been sent because of something in the character of the person sending those emails. At issue is something about their *intentions*, which are understandable even if they may merit rebuke.

When one colleague complains to another that he or she has sent too many emails, he is not saying "Use only five emails because ten are too many." He is suggesting that the person sending too many emails (and hence using too many words) knows *already* what the right number of emails ought to be and that the real question is why they are ignoring it. In some situations, people may not know the correct number of emails (or some other mode of communication) because they are learning. If so, they will be taught, and their mistake corrected. These instances aside, the complaining individual thinks not that a colleague has made a mistake but that the error in question is intentional. There is *reason* for this behavior.

So when they complain, what are they doing? Are they saying, indirectly, "Send me the correct number of emails"? I think that they are seeking a different sort of explanation. They want to know whether their colleagues are not concentrating on sending the right number of emails and hence not focusing on the job at hand because something is on their mind, for example. In a different case, they want to find out whether their colleagues' behavior is deliberately motivated by a desire to distract them. They are thinking that their colleagues are not communicating efficiently because of something in their motives. In either case, then, the manifest behavior—sending too much or even too little email—is an indication of something about the goals of the person.

The phrase "the right number of words" is misleading in the sense that it implies that communication might be a technical art with an optimum level of emails or words. In some instances, this will undoubtedly be the case (although these instances might be peculiar and hence indicative of the different contexts I am focusing on). Some very constrained registrars of communication have the humanness extracted from them, and the humans are acting (as best they can) as automatons or as proxies for machines. Air traffic control comes to mind. Pilots and controllers simply share statements of facts and declarations of intentions and are terse in the extreme. But in less regulated everyday contexts, there is a level of communication that is appropriate, and it applies in almost every context where one can think of communicating occurring. Not everyone is obliged to speak to a certain extent, but there is a presumption that there is a proper length or form, and this presumption allows people to notice when, for example, someone goes on too much or when they say too little. The words *taciturn* and *prolix* function because there is some kind of commonsense measure of length in communication. In all communication, from ordinary face-to-face conversations to the most elaborately constrained forms of being on touch, there is a presumption of economy—that participants say what is needed but not more than is needed or that they do not say less than is needed. This sense of economy is also a related to what is thought to be elegant, decorous, eloquent, and charming.

Whatever the actual details of this code of conduct, because the essential component of the human act of communication is a moral valence, this economy will therefore have moral structures. The philosopher Grice seeks to examine these structures in his *Studies in the Way of Words*, and they give human contact its peculiar resonances. These resonances are intentions and lead us to recognize that asking things like "Why does someone go on?" or "Why does someone keep quiet on the subject?" is perfectly commonplace.

The intentions that these queries point to are not abstract or obscure. They are not things that people have difficulty in ascertaining. Take the example of the person who wonders whether a colleague has something on his or her mind. Here the one judging might posit the following sorts of reasons—that the colleague has had a row at home and that this has unsettled him or her. In the second instance, where someone seems to be deliberately distracting a colleague by sending too many emails, the kind of understanding one can imagine being applied might entail reflecting on what is known about the colleague's character. Perhaps the colleague is easily bored, for example, and so might seek mischief as a way of slaking that boredom. In both cases, the knowledge is simply everyday knowledge about human nature and human character.

Such judgments might be prone to error, and people will have different abilities in this regard. But such judgments typically *assume shared knowledge* of what is elegant, economic, appropriate, or eloquent. These words all allude to same basic property that allows the judging persons to scope the frame of their judgments. Where those assumptions about economy don't apply, other kinds of judgments will be invoked, and they often are much less severe—less judging. A person who does not know how to do something is less culpable for failing to do that something than one who does. But for those who share understanding of the communications in question and of the doings they are entwined with (love, work, play), then a failure to abide by that understanding is used to indicate something else. Here terms like *mischief, laziness, resentment*, and *distraction* come to be applied.

This is important as we move toward making judgments about communications overload. Measures that change the communication act to something other than efforts to create, sustain, violate, or alter ties between people (creating society) will probably remove an important dimension from those acts—a sense of the *performing human* in those acts, the one who is judged when the judgment about the act is made, the *who* of the communication act who knows (somehow) what is a good performance and what is a poor one. I am not proposing that one can judge a communication only when one knows who the communicating person is. When a communication is sent by a stranger or when it is sent anonymously, the absence of that source information is itself used to judge the acts in question. We always start from an assumption that the who matters, even when we don't know who that who is. And we also start from an assumption that we communicate with people who already know what the communication in question ought to be about and that they can then decide to create mischief, trouble, or play by doing otherwise.

And this leads us to the rub. The preceding discussions presuppose that the participants in some setting (some habitus) know what is appropriate and what is not and can discern the forms of relationship that the people they are communicating with have with them. But this vision of the world is accurate in some instances and not for others. The world fabricated through communication might have patterns in which such judgments and measurings can be made, but it also consists of settings where things are more complex. Although one might agree with Grice that in everyday conversations, people normally say enough to allow our meanings to make sense but don't offer more than might be required (for fear of being seen as long winded). This doesn't mean that all contexts have the same rules of thumb, and it doesn't mean that applying these rules is straightforward in all cases. Passing the time away through chatting in a fashion that keeps a balance in who says what is one thing, but knowing what is enough at work is another. Part of the problem of work communication, much of which is through email, is that it is unclear what *economy of expression* might mean. Judgment about that can best be made when the topic or purpose of a communication is clear. But emails are often sent on subjects that recipients know little about and from persons whose relationship with them might not be easy to discern.

We have all experienced moments when a distant colleague we have never heard of communicates to us. We wonder how they got our name and whether the manner in which it was given obliges us to behave as if they do know us. If a close colleague suggested they contact us, then we might feel that a response is required in honor of that relationship, even if the one who has made the contact is a stranger. The moral implications of sociality can operate indirectly. And it might also be that the topic of the communication obliges you. Your job might require you to know something about such and such, and the stranger might simply be exercising his or her rights as a fellow member of the organization whose request is a symptom of a division of labor. "It is your job to know this. Please share it with me," one can hear the sender thinking. Failure to respond to this sort of request would violate another kind of tie—the one of organizational compliance (see Bittner 1967). But sometimes interpretations that allow one to make sense of a communication are not possible. One might not be able to discern how someone got our name or grasp the point of the subject conveyed in the message. Sometimes emails do seem pointless.

If this is the case in work contexts, then it is surely more likely to be the case in new contexts where the purposes of the communication are not clear or have not settled into a routine and where the relationships between the communicants are obscure. Many social networking sites create such confusions. One can imagine users faltering between thinking that a site is work-related, a mutual-interest one, or a social one (a site for making friendships). They stumble when they get their categorization wrong. When they do, they judge their acts of communication and the acts of others wrongly. Although knowing what is entailed in any act of communication turns around assumptions that are commonplace (who, what, why), these assumptions cannot always be applied. The seams that one ordinarily relies on suddenly don't do the work we expect, and instead of seeing the man in the suit, we see pieces of fabric unlinked and scattered—the world without form.

CONCLUSION: FROM ASSUMPTIONS TO JUDGMENTS

It is intrinsic to human nature to judge communication, and those judgments are about a range of issues—from whether one knows the person communicating to what the topic of the communication act might be. The ability to judge presupposes the possibility—indeed, the strong likelihood—that those judgments will err in some way. They may be fumbled at and wrong because the people who make them and the things they are about are complex and subtle. The judgments in question may be easy to make, but they are rarely certain, and they may be subject to review, too. Judgments can be about character and the relationships that people are in, they can be about perceptions of appropriateness (or economy), or they may be about the topic or function of an expression. All of this is related to rights to communicate and their corollary, the obligation to communicate given the social connections that people have or want to make.

All of these criteria are intimate properties of the texture of communication—the weave that binds us in a fabric of interconnection that is mediated through our acts of communication. The consequences of our expressions—their meaning and moral implicativeness (what it implies about next acts)—are not freestanding and singular acts. They are located within the expressive bonds that have constituted the relationships in question in the past, present, and future (in terms of prospective obligations that ensue after some act). And each act needs to be understood within a larger ecology where each act has its own relational meaning and where its history, present and future, is understood in reference to what other acts, undertaken at a similar time, achieved themselves. Each type of communication act is undertaken through a mode, and each is to be understood in terms of what that mode allows that other modes don't. This as much anything else—economy, familiarity, ease of use—makes the selection of any one type of act tactical.

As noted in chapter 5, the ability to manufacture and sustain this texture has as much to do with the articulation of words as it has to do with the articulation of the body, as much to do with an adroit fitting of embodied meaning in particular technologically (or otherwise) enabled places and times as it has to do with the management of, say, friendship. As we turn to the task of asking what we should seek to attain with our communication acts in a communications-rich landscape, we need to acknowledge how communication acts are wrapped up in the intricacies of human affairs and how those acts are constitutive of those affairs. We are both as individuals and as communities the products of our communication acts. The new technologies that we are transfixed by (and the new ones we are inventing) will themselves come to be wrapped up in those intricacies, in ourselves, and in the societies that we make—although the ways that this is so might not be certain or obvious at first. We might want to talk about volume and overload when we think of being in touch and the burdens it seems to impose on us, but we can't point to simple or certain measures. We need to allude to the vast range of human performance and social differences that constitute both the grounds for and the purpose of communication acts. Keeping in touch is a beguiling phrase in the twenty-first century, when we have invented many new ways of doing it. Although we might be confronting a new landscape, we approach it knowing a great deal about what communication entails. We don't need to ask why or what too much means. We need to focus on where, when, how, and who and how these categories are to be applied. We have the wherewithal. We just need to use it.

NOTES

1. For other inquiries, they may well be legitimate physiological questions.

2. For discussion of this point, see Norman Malcolm's *Memory and Mind* (1977). For the current concern with the design of devices that can augment memory and how this muddles up the communicative acts of memory with the scientific, see the paper I coauthored with several colleagues, "The Past Is a Different Place: They Do Things Differently There" (2008). There is a bigger issue here, though. As Oswald Hanfling notes in *Philosophy and Ordinary Language: The Bent and Genius of Our Tongue* (2000), words or phrases have a kind of grammar of meaning to them that can imply something that is not said expressly but is conveyed, and one of those ideas is that memory is a thing or a place. Hanfling makes it clear that this idea (of memory as a thing or place) is better thought of as a grammatical manifestation that leads to confusion. Memory is an important conceptual category, but the relationship between language and doings is not as it might appear, and we should be careful not to be tricked by the grammar of the words we use when evoking memory.

3. See, for example, the curious claims made by Alex Pentland in his *Honest Signals: How They Shape Our World* (2008). He argues that deep signals are embedded in communications that reveal real intentions

and meaning. In this view, the uniquely human aspects of communication (that it is mediated by language and embedded in a society or a culture that is manufactured through that communication) matter less to him than do honest biology-based signals. These, he believes, were developed when the human species were primates. Such an argument discounts the ubiquity and diversity of effort that is put into organizing human affairs, efforts that are essentially constructed through communication acts. Honest signals are one element of some acts, but it is not likely that they will be relevant for many of those acts. Society is remarkable for its diversity, and this reflects and is constituted by the diversity of the communication acts that make it. Society doesn't turn on a distinction between honest signals or dishonest ones. After all, when one is having a conversation, one doesn't doubt the honesty of the other conversant, even if one cannot see, feel, or sense the primate signals that they might be conveying (how would such signals travel via instant messaging, for example?). As Grice notes in *Studies in the Way of Words* (1989), such signals (the honest ones) have nothing to do with the act of chatting. Chatting is about passing the time, not agreeing something is true. If Pentland was right, society would be akin to a society of autistics—chronically collapsing when trust comes to be in doubt. Our society doesn't function in this way. It isn't based on honest signals.

4. In anthropology, talk has never attained the centrality that it has in sociology. Instead, anthropologists view *text*, the written word, as the organizing agent of society. This manifests itself in the self-regard that anthropologists give to their own texts and to the use of the written word in contemporary forms of mediation. This viewpoint seems to derive in large part from James Clifford and George Marcus's *Writing Culture: The Poetics and the Politics of Ethnography* (1986). Regarding the view on anthropologists' texts, see, for example, David Mosse, "Anti-Social Anthropology? Objectivity, Objection and the Ethnography of Public Policy and Professional Communities" (2006). See also Adam Reed, "'My Blog Is Me': Texts and Persons in UK Online Journal and Culture (and Anthropology)" (2005). Numerous monographs emphasize the written word and its technological mediation. See, for

instance, Jenny Sundén *Material Virtualities: Approaching Online Textual Embodiment* (2003), which is driven largely from a literary theory point of view but is nevertheless anthropological in its concerns. In contrast, sociology has always treated text as offering various instantiations of Garfinkel's basic point that language (words, in this case) at once describe what it also organizes.

5. For a good overview, see Corinna di Gennaro and William Dutton's "Reconfiguring Friendships: Social Relationships and the Internet"(2007), where they note that only about 20 percent of those who use the Internet end up extending (reconfiguring) their social networks—making new friends.

REFERENCES

Austin, J. L. 1962. *How to Do Things with Words*. Oxford: Oxford University Press.

Bittner, E. 1967. The concept of organization. In *Ethnomethodology*, ed. R. Turner. London: Penguin.

Blackburn, S. 1984. *Spreading the Word: Groundings in the Philosophy of Language*. Oxford: Oxford University Press.

Bourdieu, P. 1972. *Outline of a Theory of Practice*. Trans. R. Nice. London: Cambridge University Press (translated 1977).

Bourdieu, P. 1984. *Distinction: A Social Critique of the Judgment of Taste. Trans. R. Nice*. Cambridge: Harvard University Press.

Cavell, S. 1994. *A Pitch of Philosophy: Autobiographical Exercises*. Cambridge: Harvard University Press.

Clifford, J., and G. Marcus. 1986. *Writing Culture: The Poetics and the Politics of Ethnography*. Berkeley: University of California Press.

Diamond, C. 2008. The difficulty of reality and the difficulty of philosophy. In S. Cavell, C. Diamond, J. McDowell, I. Hacking, and C. Wolfe, eds., *Philosophy and Animal Life* (43–90). New York: Columbia University Press.

Di Gennaro, C., and W. Dutton. 2007. Reconfiguring friendships: Social relationships and the Internet. *Information, Communication, and Society* 10(5): 591–618.

Dourish, P. 2001. *Where the Action Is: The Foundations of Embodied Interaction*. Cambridge: MIT Press.

Garfinkel, H. 1967. *Studies in Ethnomethodology*. New York: Prentice Hall.

Grice, P. 1989. *Studies in the Way of Words*. Boston: Harvard University Press.

Hacker, P. M. S. 2007. *Human Nature: The Categorical Framework*. Oxford: Blackwell.

Hanfling, O. 2000. *Philosophy and Ordinary Language: The Bent and Genius of Our Tongue*. London: Routledge.

Harper, R., D. Randall, N. Smyth, C. Evans, L. Heledd, and R. Moore. 2008. The past is a different place: They do things differently there. In *Proceedings of DIS (Designing Interactive Systems)* (271–280). New York: ACM Press.

Harris, R. 1981. *The Language Myth*. London: Duckworth.

Jefferson, G., and E. Schegloff, eds. 1992. *Sacks: Lectures on Conversations*. Oxford: Blackwell.

Locke, J. 1690. *An Essay Concerning Human Understanding*. London: Penguin (reprinted 1997).

Malcolm, N. 1977. *Memory and Mind*. Ithaca, NY: Cornell University Press.

Mosse, D. 2006. Anti-social anthropology? Objectivity, objection and the ethnography of public policy and professional communities. *Journal of the Royal Anthropological Institute* 12(4): 935–956.

Nietzsche, F. 1961. *Thus Spoke Zarathustra, Trans. R. J. Hollingdale*. London: Penguin.

Pentland, A. 2008. *Honest Signals: How They Shape Our World.* Cambridge, MA: MIT Press.

Plato. 1999. *Phaedrus.* Trans. B. Jowett. Champaign, IL: Project Guttenburg. Available from http://www.gutenburg.org/etext/163.

Reed, A. 2005. "My blog is me": Texts and persons in UK online journal and culture (and anthropology). *Ethnos* 70 (2): 220–242.

Searle, J. 1969. *Speech Acts.* Cambridge: Cambridge University Press.

Sundén, J. 2003. *Material Virtualities: Approaching Online Textual Embodiment.* New York: Peter Lang.

Wittgenstein, L. 1953. *Philosophical Investigations.* Trans A. N. Anscombe. Oxford: Blackwell.

Zuboff, S. 1988. *In the Age of the Smart Machine: The Future of Work and Power.* New York: Basic Books.

7 THE TEXTURE OF AN EXPRESSIVE FUTURE

PREAMBLE

Over the past few years, I have been involved in research looking at older people's communicative practices. As represented in the gerontology publications, older people are viewed as often suffering from a physical decline that inhibits their capacity to communicate. In the sociological literature, older people suffer from an aversion to new technologies, which has the same consequence as bodily decline: people become *out of touch.* My research has sometimes been commissioned to remedy these concerns—either by using novel human factors and ergonomics tricks to design interfaces and hardware that are easy to use for those with declining dexterity of the hand or by offering digital communication experiences that are disguised as something else that old people might be sympathetic to (such as forms of game play). Older folks can't or won't communicate, this view holds, and this needs correcting. Yet each time I undertake such studies, I am confronted with the fact that older people do communicate, do so with passion and delight, and often ignore any problems with manual dexterity or visual acuity they might have. I start the research having been led to believe that "Older people are too old to learn new tricks of communication" and "Older folks can't press the buttons on mobile phones, let alone see the numbers on the screen," and then I find that they engage with the technologies of communication with delight.

People's bodies do decline, and some older people are averse to expression and being in touch. But my research has taught me that older people should not be assumed to be beyond the pale of communication. Indeed, after moving beyond these fixed views and observing what most older folks do, three main things have been made clear to my colleagues and me (Lindley, Harper, and Sellen 2009).

First, many older people assign special status to communication that is focused, is intensive, and honors those being communicated with. Certain forms of communication, such as the letter, seem to satisfy their requirements. Letters demand diligence and concentration, they explain. Writing can force the writer to pause, reflect, and get it just so. When people write a letter, they have to calm themselves down and devote themselves to thinking about what they want to say to another person. This effort produces communications that are interesting and worth something—*worth* in the sense that it cost something to produce (in time and concentration) and in the sense of honoring the person who receives it. "A letter makes someone feel special," as one individual put it.

The second main lesson that came out of our research is that many older people express disdain and dismay at what they call the ephemeral and ubiquitous messaging of the young. In their view, posting information about what one is doing on a social networking site and declaring one's status on an instant messaging client seem to suggest an indifference to those being communicated with. In their eyes, the willingness of the young to message to multiple people suggests a lack of concern or genuine interest in any one of those being communicated with.

A third lesson has to do with how older people account for communication. As those we interviewed explained, the importance of communication and their dismay at younger people's multicasting made it clear that they didn't think there is any absolute right or wrong to the behavior in question. They thought that how the young behaved is not as they would do it, but they accounted for this by saying that values have changed. "Today everyone is in a rush," they said, and this might be the reason why "No one can be bothered with communicating properly."

Older people's attitudes toward communication—their understandings of what communication is about and what it affords—are indicative of the view that I have outlined in previous chapters. Older people don't think of communicating as a telementation task. They view communicating as partly an exchange of stuff—news, comment, tales, and gossip—but essentially a performative act that often has moral implications. In their view, communication is a task that honors those involved. One communicates to celebrate and cement friendship and to express the values that one holds dear. These values entail making judgments about how to savor and honor friends, values that older people think are different in different generations.

In these respects, older people have what I think is the right understanding of communication. They aren't deceived by myths of language, odd psychological arguments, or Turing theoretic ideas that might apply to other forms of communication but do not account for the specifics of human expression. They understand human communication for what it is—a kind of moral order that is composed of expressive acts and judgments about the values of those acts bound to times, places, and embodied skills.

The cargo of the previous chapters affirms and justifies this claim. It is from this view that we can consider the question posed at the outset—why many of us complain about communication overload and yet find ourselves entranced by the use of new means of communicating. But the cargo also makes it clear that solving this conundrum is not easy. The previous chapter, for example, describes the structural elements of the texture of communication that can help clarify the things we have to assume when we ask our questions. Some aspects of the texture of communication are so basic that they are taken for granted. The web of connections made by two people when they become friends transcends time and mode of expression, for example, and hence has properties that need to be seen as the basis of telecommunicated relations, whatever form they have or whenever they occur. Saying that digital networks create new forms of social relations ignores the fact that social relations have always been made up of networks, and the distinction between those one does and does not know is a cleft in *all* networks, mediated or otherwise.

A similar difficulty applies when it comes to understanding how to measure acts of communication. To say that communications acts are not to be judged by their volume but by the moral system of which they are a part seems easy enough to accept, but this leads to the difficulty of defining what different communication acts enable if not volume. For older people, the written letter honors those that they communicate with, and they seek honor in other modes, thinking it the key property of their acts. But honoring is a starting feature of the choices that older people make, not merely an outcome that can be discovered only after looking at the behavior. Before they choose a method of communicating, older people *already* know what honoring means and how it can be delivered through acts of communication. They discover through use that new methods enable it more or less, and they refine their understanding of the world they live within accordingly.

In the first part of this concluding chapter, I revisit how we got to understand these arguments—the mulling over of evidence and issues that I began the book with, the questions that my own research experiences engendered, the insights (and sometimes curious absences) in sociological research on the act of communication, and the fate of technologies of my own and my colleagues' devising. The second half of the chapter addresses questions relating to the value of our communication acts, our delight with new modalities that we use to undertake them and our complaints about the burden they impose on us. Explaining this delight and these complaints is done through understanding the complex judgments people make as they navigate their way through the range of possibilities that is the texture of our communications age. When we look at these judgments we might not find the answers that we hope for—not easy ones, nor ones that are easily grasped. Nor may they be ones that we would agree with ourselves. As should be clear, choosing to communicate is a means of conveying identity. People will therefore choose differently. Besides, it is not only character or identity that is at issue when we ask whether we communicate too much or why we are entranced by new channels. We are also pointing toward something essential to the human predicament. Our deeds of expression (our mediated expressions being one kind) are how we make our society and ourselves. Our complaints about the economy of our lives—about how much time we have—are measures of something much greater. They comment on the ties that bind us to each other and the ties that, in allowing us to express ourselves, make us what we are, what we are seen to be and who we are seen to be connected to.

WHAT HAVE WE LEARNED: EMPIRICAL FALLACIES

This book begins by sketching the empirical circumstances in which we live (the numbers of messages and channels that we currently exchange and exploit) and also our vision of what it means to be (how we think of ourselves). I suggest that over time we have shifted what motivates our communication practices. If there was a time (say, twenty years ago) when we would scold ourselves that we didn't communicate enough because it was too much hassle and not much fun, in the second decade of this century, we have reached almost the opposite condition and are scolding ourselves to calm down and communicate less. Twenty years ago, we were convinced (as we are now) that communication is good for you, but the levels of what was optimal had not been achieved. Today there is a feeling that we have gone beyond the right balance and we communicate too much.

How did this change occur? Why? Did technology entice us into new habits? Or have we been developing a philosophy—a morality—that says, "Keep in touch more than you used to. It's good for you." and "Being in touch is what makes the digital age different—better!"? And has this morality or credo gotten out of hand? Or has this morality interplayed with developing communications technologies to foster the world we see around us now—one that we have lost control of? Has this marriage of ideas and technologies—designed for a world where being in touch is valued—somehow produced progeny that has led us beyond an ideal point toward overload?

I ask how a concern with the values of communication and the problems it can create has produced a cultural and technological landscape where we have begun to see the world in terms of communication. It has produced a world where we have come to see the people within it—the communicants—in terms of communication. Communication seems to be the leitmotif not just of our modern society but of us too—the people of our day and age.

For example, I start my discussion with reflections on why we think that communicating is good for us. I look at the history of the written letter and the virtue that letters are somehow thought to bestow on the sender and the recipient. That they were good for you was something that had to be cultivated, developed, and sold to an otherwise indifferent populace. I look into the volumes of letters sent over the years (in the nineteenth and the twentieth centuries, for example). These volumes should be high if letters are meant to be good for you. Indeed, we imagine that our Victorian and Edwardian forebears wrote copiously to each other, spreading thereby the civilizing virtue of the written—the communicated—word. But this isn't the case. People didn't write much in the past, and they write even fewer letters now. Nevertheless, the era that produced the idea that letters are good for you (especially the nineteenth century) was the time when our current view of the value of communication started to take hold. As David

M. Henkin notes in his book *The Postal Age: The Emergence of Modern Communications in Nineteenth-Century America* (2006), the "golden age of letter writing" helped produce the metaphysics of the contemporary age of communication.

Even if we were to measure our own or prior ages in terms of numbers or volumes of communications, then what we find is that counting isn't all that easy to do, even if it turns out that one ends up with small numbers, as in the case of letters. Although the question of "How many?" might seem to be a banal quantitative question, it is not as easy to answer as it first appears. Nor is human communication as simple conceptually as might be suggested when the word *quantity* is applied to it. The term *quantity* might seem to involve counting, timekeeping, and making judgments about the allocation of human efforts and resources. And whatever else it might be, communication is this—human action. But determining what statistics might be relevant when making judgments about that action is not straightforward.

If the question is whether we communicate too much, one issue to navigate around is what the communication act is. When I type instructions into a computer, am I communicating? If I use spoken words when I do so, does that make the act the same as when I speak to a person? Some researchers, like Clifford Nass and Scott Brave in their *Wired for Speech: How Voice Activates and Advances the Human-Computer Relationship* (2005), think so. But the two acts are not the same, for fundamental reasons about the nature of human expression and nothing to do with its visible form. In this view, machines can communicate with each other with words, people can use words to send instructions to a machine, and people can use words to chat with each other. But though these three acts seem similar and all use words (or *natural language*, as computer scientists put it), the things that are achieved in these cases are fundamentally different. Two are forms of data exchange, and the third (when people talk to each other) can be merely a case of people listening to each other—hardly the same at all.[1]

If we confine ourselves only to human-to-human expression, we still have problems of measuring. If we start thinking of letters, for instance, things might sound straightforward. But all letters? Some letters are merely orders or receipts, while others are communicating something of personal value that one might want to attend to. Communications aren't all of a muchness, even in the same medium. Besides, are letters and other types of communication always singular and easy to measure? When we email, we often are doing other things, such as responding to a written letter. But we might also be instant messaging. When there is a concurrency of acts, separating them all out into neat categories might seem to be required. Unfortunately, this can be very hard to do. Task measurers have great difficulty separating tasks. Concurrency of actions is a big a problem for them. "If only people would do one thing at a time, we could produce better data," one can hear these measurers muttering. But I ask whether this ideal vision misses the point, given that the overall texture of our lives combines face-to-face encounters, telephone talk, instant messaging, watching TV, and tweeting. This texture is what we need to grasp when we measure. After we make decisions about what set of things we measure, then we still need to make a judgment about whether we communicate too much. If we say, "We do it too much," then by what criteria are we judging our communications? This leads us back to reflect again on the things we measure—whether corporeal things, countable objects and actions, or ideas, hopes, and aspirations. These are not the same, even if they are bound to each other. Ideas lead to actions, after all, just as actions lead to ideas.

As we construct our measured visions of how communications affect things, we also construct pictures of what we think the human is—of who or even what we are. We might start off measuring emails, but we end up making comments about the human as a type of creature that has a "ghost" inside of it wanting to get out via communication, a view that John Durham Peters explores in his *Speaking into the Air: A History of the Idea of Communication* (1999). Or we start off with measures of human communication—of how many and between whom—and end up saying that what matters is not the human but the network, as does Manuel Castells in his *The Internet Galaxy: Reflections on the Internet, Business, and Society* (2001). In short, we think we are asking a simple question or set of questions (why do we communicate, how much, and is it good for us?), but we discover that we are trying to answer a whole range of questions, even some having to do with the essence of the human condition.

In chapter 3, I describe the vision (or set of visions) of human communicators and their communication acts that frames the design of new communications technologies. I don't offer a comprehensive history of technological innovation but explore what have been the key visions or principal goals that drive the communications technology inventions that my colleagues and I devise. My own professional experiences—journeys through Xerox, academe, start-ups, and Microsoft Research—are the resource here. Many of the things that my colleagues and I build (irrespective of whether we have made money on them) are fairly typical of the inventive efforts that have appeared over the past fifteen or twenty years. My main concern is less with the technologies and more with the ways that these technologies are constructed on the back of a model of humans and their communications practices that emphasized the *body mechanics* of human communication at the expense of other dimensions, particularly the performative. This model or vision provides a common ground on which shared ideas and design possibilities can be built. This limited, body-emphasizing view gives us something to grasp onto in our imaginations. Our inventiveness does not intend to be philosophy, after all. Nevertheless, our visions of the human and of the human properties that matter for communications technologies lead my colleagues and me along a particular path—where more is viewed as better and to a move from the asynchronous to the synchronous, from the auditory to the visual, and from the unisensual to the multisensual.

The systems produced as a result of this view of the communicating human offer much less sensual richness than is aspired to. One of the by-products of this inventive landscape is the use of textually mediated forms of communication. Many of the current (and hugely popular) textual-mediating applications (like instant messaging and Internet relay chat rooms) are distant by-products of attempts to build systems that offer more than words. Indeed, conveying words seems to be the greatest legacy of these inventive efforts, despite the hopes of the inventors (including me).

The vision that underscores the inventive landscape has its roots and main compass in the intellectual tradition that produced computer science. This discipline takes for granted the idea that mathematically based knowledge (about the human, for example) has a long way to go before its limits are reached and that those working in the fields are remarkably ambitious. Like the mathematical philosopher Roger Penrose (1989 and 1994), most computer scientists believe they can and eventually will design systems for almost anything that people would want to do.[2]

In having these aspirations, these individuals presume a corollary. Although mathematics can let them do almost anything, they are convinced that humans *can't do everything*. In their view, humans are limited machines (albeit remarkable biological ones) and have a threshold beyond which they cannot go. This assumption is commonly held in disciplines that are closely allied to computer science (such as cybernetics) and in recent derivatives (such as cognitive science). In all these disciplines, the individual human is thought to be merely a biological processor of information—little more than an enormously complex computer made in flesh and blood.

Although these arguments and corollaries can be pragmatically helpful in the business of invention and useful in various types of sciences seeking to explore the physiology of the human, they don't account for the rich properties of human communication. The vision used to imagine new scenarios of technological mediation simply don't seem apposite for the nature of that communication. The vision used to orient design is of a world that is not the same as the one real people populate. The epistle sent between two persons can create a special bond between the correspondents and somehow lets them transcend time and space in their relationship. How a letter can do this is subtle and almost miraculous, but it is understood (somehow) by all who have crafted or received one. In contrast, many of those in the world of invention have difficulty distinguishing between an absence in a communication, a void, and a chosen silence that has meaning. One is a nothing, and the other is a moral act, a snub, an expression of disdain.

A SOCIOLOGY OF EXPRESSION

The human in the world of invention, then, seems to be a remarkably impoverished creature that is devoid of the capacity to make even the most elemental communication acts—to reply to a hello with a silence. Chapter 4 presents an alternative, richer vision with evidence found in sociological studies. This vision is confined to what is found in the sociology of mobile telephony. The work of Barry Wellman is considered especially carefully, since his views about the changing patterns of society that mobile phones and Internet-enabled communications are creating are representative of the typical view from sociology. This view emphasizes change in social structure, especially those to do with the shifting importance of geography in social relations. Once geography was the essential foundation of those relations, this view holds, but now newer forms of social relations reduce that centrality. Wellman, like most sociologists, assumes that there is a move away from fixed-place societies toward an information-based connectivity, and they assume that such new types of connections weaken (or even obviate) geographic ties.

However, in the empirical investigations reported in the chapter, we find that the posited shift between a world of fixed location to a world of spaceless social relations is exaggerated and disregards a wealth of complex social relations in which geography and spacelessness are only two dimensions in a heterogeneous grid of social structures. Some people use the virtual address book on their mobile phones as a pretext to interact with others when they are together (to show and share their collections of contacts during a conversation, for example). And they use the same technological apparatus (a mobile with a virtual address book) to force others to contact them remotely when they are apart. When remote access to them is requested via their mobile (when a call is made), they can see who the caller is and can ignore that call (via call forwarding, for example) without the caller knowing. This lets them avoid conversations that they don't want. Callers are sometimes willing to go along with these schemes, especially if they want to avoid an embarrassing conversation.

Studies on text gifting, meanwhile, demonstrate how the technically simple properties of short message service (SMS) are leveraged by users to afford the age-old practice of creating social bonds—but in ways that cannot be thought of in terms of geographic distances between bodies. When it comes to understanding human communication, the geographic focus of many sociologists are too limited, placing too much emphasis on space and not enough on acts of communication.[3]

Social relations, as encompassed by various forms of communication acts, are rich and diverse, changing in their details sometimes more than in their general character but changing nevertheless in ways that are bound up with their delicate and diverse forms. The research focused on in this chapter only deal with one channel or medium, the mobile, but lessons derived from this research apply to the use of Internet-based interactions (via instant messaging, Internet relay rooms, social networking sites, blogs, and so on). As Corinna di Gennaro and William Dutton remark in their paper "Reconfiguring Friendships: Social Relationships and the Internet" (2007), the use of the Internet to create new relationships is only one of various modes of creating relationships, and it is used by only about 20 percent of Internet users. Its rich, artful, and diverse forms make it difficult for standard sociological categorizations of the human (age, gender, education, income, and so on) to explain that richness. As di Gennaro and Dutton put it, "The dynamics of online relationships are driven more by idiosyncratic digital choices made by users of the Internet than by any mechanistic social . . . determinism" (2007, 591).

OUR OWN TECHNOLOGIES

All these arguments provide the background to chapter 5 which reports on the use of two technologies of my own and my colleagues devising. Glancephones and the Whereabouts clock are technologies developed to offer solutions to what we perceive to be need (and thus are in accord with visions that emphasize the body and the mechanical properties of human expression), but in their deployment, something forces us to alter and develop our views. These devices do not so much solve a problem as create an opportunity for altering what people do, want, and hope to show about themselves. These doings and desires are bound up with the enjoyment that people find in expressing themselves and telling their stories. The richer, more nuanced view that my colleagues and I develop as a result of our research studies with these technologies reflect the types of idiosyncratic choices that di Gennaro and Dutton indicate are important but don't describe in their own work, presumably for methodological reasons (because the data wouldn't let them—i.e., age, gender, income, education)—namely to do with the *who* behind the communications act, the *what* of the particular relationships that can be cultivated through expression, and where these acts are carried out (at work, home or even leisure settings, for example). Ties of friendship can be made through mockery, for example, and family ties can be strengthened through acts that show affection (the making of tea). We do not predict these uses beforehand. These new modes of expression and new communications acts evolve through use. These doings and desires are bound up with the enjoyment that people find in expressing themselves and telling their stories.

Telling the story of how our own research leads us to look beyond the landscape of invention toward the large question of communications overload. We shift our understanding of communication to include more social and moral concerns than were required in the past, and we develop a view that lets us address why we and those we study complain about the communications burden yet desire to express ourselves more (even as we complain). Both the users' and our own values are at the heart of this concern, and these values are the common ground of humanness in question. What is required are answers that are essentially moral, and judgments about the value of communication acts ought to be in terms of what they achieve as social actions with moral implications. Communication acts are not to be thought of as, say, a transfer of information between two machines (in the form of human bodies, perhaps) but as acts that alter the moral fabric of the relationship between the senders and the receivers. Sometimes that alteration means that one person has some information that he or she did not have before, and sometimes it means simply that one has been courteous in listening to the other. In either case, it also means that the relationship between sender and recipient has altered, too. These kinds of measures are related to the *performative* values of each and every communication act. For us to understand and judge these values, we need a sensitivity for the performative aspects of them to be at the core of our analysis.

Such sensitivity is not easily cultivated. In chapter 3 we see that the business of invention is constructed around a vision of the human that is good for inventiveness (if by that is meant numbers of patents filed and technologies developed) but it is a vision that emphasizes the body at the expense of a richer view that encompasses intention and mindfulness. This contrast between the body and the mind is itself difficult to steer around. Oftentimes attempts to correct the body emphasizing view lead to an excessive interest in the mind—sociology can suffer from this, for example. Bourdieu teaches that a better view balances a concern for the body and the mind, but adds to this recognition that the place in which the human acts is also important. The habitus of communication acts are the locales in which embodied and mindful behavior have their proper place and function; it is where they are developed and how they are judged; when they are undertaken out of place their value is hence altered.

Key to the sensitivity is recognition that communication acts are to be understood in three dimensions—the how of the act itself (the bodily skills used in their performance), the who of the act (where one needs to be alert to the intentions of the actor themselves and how the undertaking of some act conveys a sense of identity or self for that person and to the audience or recipient of that act) and, third, the where of the act (the location of its performance). Though one might imagine that an act of communication might be undertaken anywhere, anytime, in practice where and when it is undertaken colors the meaning given to it and the judgments about the performative value it might have. All acts are bound to place just as they are bound to identity just as they are to when they are undertaken. A lack of clarity about any of these indexes is also used as an index of meaning too. A message sent anonymously has a different meaning to one sent from a close friend, just as a passionate message sent from work is interpreted differently to one where the location of the sender is ambiguous.

JUDGING FOR OURSELVES

We have been on quite a voyage since this book's opening gambit, where I said that we communicate too much and yet seem to delight in the experiences that new channels of communication afford. But the question that underlies this gambit is how to judge and evaluate our communication acts. We have looked at what judgments on this topic might look like, what they might presume as well as enable, and how communication technologies facilitate new acts but presuppose some already existing forms of being in touch. The point of an act of communication can be many things. Even a fib can either repel or cohere and have many consequences. Fibbing attests to how diverse and peculiar acts of communication can be in their form and result.

The book opened by fielding two interpretations of the basic conundrum at its heart—whether our passion for communication is leading us toward a dystopia where too much communication obfuscates necessary expression and whether our delight in communication is leading us to refashion what we think of as our essential selves.

The first of these—the tension between having something to say and consuming time and effort in the saying—cannot be measured quantitatively in terms of the arithmetic of time versus content. It needs to be studied by qualitative judgments of another order. The salient property of a communication is not the stuff contained in the communication (the words, sounds, or other type of content). It is how that stuff (whatever it might be) allows certain kinds of things to be done by the people involved in the communication exchange. Communication between people is a performance that ties people together (or throws them apart) in various ways.

People communicate to make a society. (Sometimes that involves an exchange, but the point of their exchange is what is at issue.) Society is a complex object. People's acts of communication spread through time as well as space, for example, reflect and embody matters of personal decorum and dignity, sustain notions of interpersonal ritual, and produce ties of romantic delight. They can even be a vehicle for memory. Having a sense of the past is key to having a sense of oneself in the present, and many of these memories are constituted and evoked by our acts of communication. The texture of relations created through communication acts is consequently rich and subtle, startling and commonplace. Given this, new modes of communication will add to this richness and extend the weave of communication texture that binds people together into new social forms.[4]

Fashioning society is not all that communication acts do, and this leads us to the second basic question—what we are. When people act or perform their communications, they are making ties between themselves and others and they are also making themselves—portraying and evoking their character. When we judge communication acts, we aren't judging only the social nexus that ensues or is reflected in those acts. We are also judging the persons doing the performing. Judgments of this kind are (or ought to be) skillful and compassionate and should distinguish between social typologies of character (stereotypes) and accurate, rich descriptions of uniqueness in human nature. (But it must be noted that typologies of various kinds are a key tool in this work.)[5] The judgments are made subtle and complex by the subtle and complex ways in which character can be portrayed through various communications channels. In some channels, character is made conspicuous and central, and in others, it is deliberately obscure, almost lost from view. Some channels let character have a more constructed, fictional air, others less so. People are how they express. The richness of their expressions and their adroitness of articulation are measures of who they are. They are the ways in which they are themselves.

Our goals have been to understand how we can judge carefully and delicately what is excess and redundant, what is appropriate and apposite, how we might judge each other through the prism of our communication acts, and what things cannot be inquired into through that prism. The riches of the communication act can be understood only by knowing the purposes and consequences of communication, our presumptions about ourselves when we express, our additions to that sum of understanding when we communicate anew, and the vast nuances of connection that are made possible each time we call, text, post, email, IM, blog, or tweet.

DIGITALLY MEDIATED SOCIALITY

So we should not rush to apply metrics to the new communications channels. Some metrics might seem apposite but might mislead. Others might seem orthogonal but might offer better traction for the queries we want to answer. When we talk about riches, we need to stop thinking about counting. Wealth has many forms.

Teenagers and older young people spend a lot of time on social networking sites. But to measure this use in terms of time can be misleading. The salient consequences of using such sites might be better understood by measuring in other ways. When young people use Facebook, they might be doing so instead of hanging out on a street corner, for example (boyd 2008). In both contexts, the street corner and Facebook, they are being convivial (Shirky 2008), and they are communicating. In terms of time given to communication, young people's world hasn't changed. What once was talked about on street corners is now talked about via postings on social networking sites. The world has changed, but what that change might be and how that change can be measured might not be as we assume. We might need to look in unexpected places, and the changes might not be what we thought.

Another source of difficulty has to do with using concepts that can entangle one, and geographic metaphors are especially troublesome. Services offered by social networking sites like Facebook might be said to remove space from social relations. Teenagers can now meet anyone on their virtual street corner. But this is not what teenagers do in their accounts. Most teenagers' use Facebook to communicate with those they know by living near them—their buddies down the road and mates at school. They may have one or two more distant friends (such as cousins and people they might have met on holiday), but their buddies and mates are mostly people they live near. Geography still matters in this regard.

This means that social networking sites are not necessarily used to venture into the spaceless horizons of the virtual world to meet strangers. Social networking sites are often about deepening and sustaining rations that space has already delivered the users to. Within the confines of space, there are divides that aren't especially geographic. Kids tend to hang out with kids of a similar economic, religious, and ethnic background, for example, even if they make their selections from the neighborhood (boyd 2008). This doesn't mean that social networking sites aren't about meeting strangers. They can be and in some cases are expressly so. Online dating services are a form of social networking, even if they don't accord with some definitions that emphasize the networked public dimension. In some countries, such as the United Kingdom, Germany, and France, such sites are the third most popular technique used to meet for romance after other techniques, such as being introduced by friends, going to a pub, or clubbing (Nielson NetRatings 2005; see also Churchill and Goodman 2008).

THE DIGITAL SCALE OF SOCIALITY

That this is so draws attention to another misleading metaphor that is bandied about in discussions of social networking—that they are *social* networks. Social networking sites are used in many ways, and meeting strangers—ambling with one's digital fingers beyond the pale of one's normal orbit—is only one of them. There is another side to social networking that is easy to neglect because excessive weight has been given to the idea that such sites are instruments to meet and *only* to meet. Facebook members can easily exclude people from their accounts, and teenagers are keen to exclude certain people—not strangers but their mothers and fathers. The great appeal of Facebook is that teenagers can withdraw to their bedrooms and yet go to a virtual space that their parents cannot access. Facebook is a type of social network that allows its members to fabricate walls that transcend the weakness of real walls, although it allows people to make requests for access. Parents often insist on having access rights to their children's Faceboook accounts, hoping that this might make the children safer. When this happens, most teenagers either limit their parent's access or simply open up another account under a pseudonym that they give to their friends but not to their parents. One of the most important aspects of these sites is that users don't intend to create new relations through them but instead intend *to keep away people* that in other contexts they would find it difficult to spurn.[6] Most commentators on networks sociality don't acknowledge this point, not simply because it is about exclusion but because it emphasizes the vitality of nonmediated relationships.

One of the basic properties of social relations is the distinction between people who know each other and those who are strangers to one another. Once an introduction has been made, the existence of a set of mutual obligations to respond to a communication remain inviolate thereafter, almost in perpetuity. As we see in chapter 6, at any time after an introduction, replying to a greeting with a silence or a turning away is viewed as rude. But with social networking sites, such effrontery can be avoided, and the idea that social relations distinguish between those that one knows and those that one does not can be preserved even as it is undermined. This can be done through the management of access rights that are not made visible to the one seeking access. Here is another example of how the world is changing—not by allowing strangers to meet but by keeping apart those who already have a relationship.

PUBLIC AND PRIVATE DIGITAL NETWORKS

Another problem is often raised in connection with social networking—the problem of breaking the boundaries between the public and private aspects of an individual's life. Here there is a temptation to make gross and simple contrasts to measure these domains and make them excessively concrete and absolute. According to some commentators, social networking sites have the potential to dissolve the boundaries between these domains, and in their view, this has all sorts of consequences—mostly negative. Concerns about the breaking of this boundary via technological mediation were being expressed before the widespread use of the Internet. Joshua Meyrowitz offers one such critique of reality television shows in *No Sense of Place: The Impact of Electronic Media on Social Behavior* (1985). He claims that such shows dissolve privacy and reduce the rights of the individual. In Meyrowitz's view, the right to have a private life equates to having a right to a private space. He doesn't address the question of whether those who have more private space also have more rights or whether that private space may be better thought of as an orientation to types of behaviors rather than a geographic and inviolate zone. Meyrowitz's arguments were widely accepted in the mid-1980s and are often cited now by those analyzing social networking. One concern common to Meyrowitz's complaints about reality TV and to complaints about current social networking sites is that both can allow information that ought to remain in the private domain to slip out into the public one for all to see.

Parents worry that their kids will use social networking sites, post images on their accounts that their real friends find funny, and end up having their photos seen by the world at large. This world might find these materials offensive or might use them to hound those children for years to come. Adults can also find that their private postings have slipped out to the world at large (Shirky offers a number of examples of these in his book). But people learn from their mistakes.

When people use social networking sites, the possibility that they may be unwittingly revealed to a larger public is ever present. As boyd notes, the networked public is an imminent present and threat on even the smallest network of Facebook users. Friends can crop and post images on Flickr in a moment, and once they are there, the world as a whole can see them. But a better way of measuring social networking sites is by determining what their users intend to achieve by using them. For most teenagers, social networking services are not used to spill themselves forth in front of a scolding and hurtful public. They use these sites to keep their worlds private, small, and out of sight of parents (people who might feel that they have rights to their children's private domain). The ever present threat of disclosure is a price that such teenage users are willing to pay to secure a paradoxical privacy. It might give a frisson to the exchange of private matters, giving it an edge that makes it more exhilarating. Perhaps Facebook offers them a missing sense of danger, even though they don't fully realize the consequence of falling foul of the danger until it happens. For older users of Facebook, the threat of networked visibility has other consequences and implications.

Part of the problem for users of social networking sites and those on the outside wondering what goes on inside is that the purposes for such sites are not clear or fixed. One way to measure appropriate and inappropriate expression is by looking at what one already knows is appropriate and inappropriate. This might hold true for well understood types of communication acts, but what about acts that are enabled by new technological mediation and aren't clearly appropriate? Is teenage use of Facebook about exclusion (for example, keeping Mum and Dad at bay), or is it about titillation (spreading stories about school friends)? Moreover, what is the relationship between this pattern of expression and the fact that networked publics might appear in a moment's notice and suddenly start watching?

Social networking sites are evolving and settling into a variety of different functions with associated expressive modes. There are many types of social networks, and each has various domains for particular communication acts (various habitus). But many of these domains have not settled into an identifiable form. Because sites and services are all digital or virtual does not mean that they are all the same. Facebook is about youngsters intriguing and playing with each other and wanting to do so out of view of older people. Older people use it, as well, perhaps as a playful way of reliving their teenage experiences, partly through the giggling fear that public display can induce. Other social networking sites offer different expressive modes and different acts of communication. Dating sites are about the complex art of meeting potential romantic partners, LinkedIn is about the political system of owing and calling in favors on professional affairs, and Flickr is a means for sharing images that can be used to get in touch with people or to advertise one's image-capturing abilities. Many other services and sites are appearing, evolving, and gradually settling into something that is as yet still yet certain.

One problem with this state of play is that people might navigate to the expressive domains of social networking sites not knowing what to expect and hence not able to judge what is being communicated. They won't know what they ought to assume and so might misjudge what they find communicated. The ambiguity of function is not intrinsic to the medium of social networking sites, but social networking sites can afford many types of communication act, and they are unfamiliar territory. It currently is easy to misconstrue what any particular site is if one approaches it with the wrong set of expectations and assumptions. Commentators don't help in this regard by offering simplistic accounts of what all social networking sites do or by muddling the potential user of the medium on which such sites operate, the Internet, with unguarded use of generalizing terms. Some of the language used to define communication acts is befuddling, too, leading us into traps of expectation that aren't helpful. The words *social networking* imply for many the act of voyaging with the world at large, but this is not what some networking sites are for. As I have just noted, they are for keeping the mass away, keeping the world small, and confining it to geographic parameters (even if those parameters are themselves thought confining in other ways).

FORMS OF DIGITAL EXPRESSION: FROM BLOGS TO TWITTER

The question of what a site is for and how this allows people to use it and interpret the communications exchanged within it brings us back to another form of mediated communication I mention extensively in chapter 2—blogs. Blogs are said to have particular forms and properties, one of which is that the content posted on them has a special value, being passionate, vocal, and of-the-moment rather than bland, unemotional, and sterile. Blogs could be offering a new domain for public performance even if that performance has the feel of the private world—being less formal, less considered. A lot of commentators have written about blogs and about the move of the private into the public or the private persona into the public domain. Here, public personas get displaced, if not dissolved altogether, by the vitality and appeal of private ones.

Some commentators think this is more than a mere stylistic change and has implications for the order of society as a whole. The sociologist Richard Sennett in *The Fall of the Public Man* (1986, originally 1977) and *The Corrosion of Character* (1998) has argued for a number of years that society functions better when its members can distinguish between public and private personas because in this way the imperatives of public office (occupied by a public persona) aren't infected by private desires, passions, and personas. These latter drives might not be in accord with the demands of public office and in Sennett's opinion often are not. The dissolution of the boundary between the two has meant that society finds its public offices less likely to be held by those who honor society above their own personal desires. This places discussion about the value of blogs into a longer historical context. From this view, we should be measuring blogs not by how much time is spent on them or by how many people read them but in terms of the currency of the arguments and issues that are raised through them. They are the current domain of the public persona even if they have the style of private performance.

It is not clear what Sennett would say about blogs in particular, since his arguments are rather general and much of his work was developed before blogs became widespread. Other commentators, with similar views and concerns, have discussed blogs, though. The German social theorist Jurgen Habermas views the emergence of private voices on the blogosphere as an abhorrence since these are private personas in unregulated public places. Anyone can adopt a persona, irrespective of their knowledge, ability, or status, Habermas remarks, without thinking about whether anyone reads these sites. In his opinion, blogs undermine the importance of intellectuals in public life, allowing the less expert to take over (see his 2006 "Toward a United States of Europe"). One can wonder what Habermas means by intellectuals when one considers his arguments about the need for forums of rational debate unfettered by artificial constraint (as in his 1991 *The Structural Transformation of the Public Sphere*). Blogs might offer this forum, and bloggers might be considered intellectuals of the networked society. Still, one should not forget that blogs are simply a channel and that not all blogs will be used to constitute a new domain for rational discussion or the performance of private attitudes in public space. As Jill W. Rettberg notes in her book, *Blogging* (2008), blogs are used in many ways. The question is whether blogs of a particular type of communication act might create a threat to the status of intellectuals.

Blogs can allow a kind of discourse where the identity of writers and readers and their relationship with one another may not be salient or even visible in the performance of that discourse. In a sense, the blogosphere is a domain in which many of the traditional rules of sociality don't apply. When someone posts on a blog, they don't routinely expect a response in the way that one does when one sends an email or a text message. Blogs are not that order of connected discourse, but they do create ties between bloggers. Blogs create trails of arguments and narrative threads between various sites, creating a weave that binds those sites together. In so doing, the authors of those sites and their readers are entwined in a particular texture of communication act—one that engages but is somehow anonymous. It demands personal kinds of responses even if readers aren't morally culpable for those responses or culpable of neglect of a social relationship if they don't offer such responses (blog back). Silence in the blog world isn't a turn at talk. It's the end of a subject.

The ultimate question is the problem of choice and the things that it says about people who delight in new channels, those who prefer more time-honored ways and others who want to find a middle way of dabbling with new channels but not letting go of the old. Some philosophers and cultural commentators have argued that the past few decades have produced a culture of consumerism where a product is a vehicle to show *distinction*. These distinctions represent something other than the thing chosen and are especially good at conveying distinctions of wealth. The rich will choose the latest product not because it is the best at satisfying their needs but because it is expensive and only they can afford it. They buy it because only they can. Their purchasing affirms their status as a member of a wealthy elite. Commentators on this aspect of consumerism also argue that distinctions are continually being serviced by the provision of ever new products. As new products arrive, some people are the first to consume those products, and this choice (being first) is another form of distinction. The commentators go on to say that this consumerism can never cease because it is intrinsic to the nature of this cultural practice that newness must constantly be manufactured. In this view, a desire for a distinction ends up creating a consumer dystopia—a desire to create new products not because they produce any real benefit but because they satisfy the desire for newness. Newness becomes a value in its own right—newness for newness sake, distinction for distinction's sake.

This is a summary of the views of the French cultural philosopher Jean Baudrillard, for example, who wrote his *The Consumer Society: Myths and Structures* some forty years ago in 1970. Although that was long before the Internet, a similar argument could be made about our consumption of new channels of expression today. We delight in the new even if what we choose doesn't afford an effective solution to our communication needs or replace older channels with something better. As we look back over a hundred or two hundred years, we can see that even though many channels were introduced with the expectation that they would substitute and improve previous communications technologies, this didn't prove to be the case. Just as the Victorians thought that their telegraph would replace the written letter (see Tom Standage's *The Victorian Internet,* 1998), so we have seen that the Internet has failed to replace the written letter.[7] Whatever the hopes of our forebears, we have more ways of expressing ourselves and being in touch, and this would easily lead us to think that this landscape is the result of the kind of consumerism I have just described.

Some people use new channels because they are new, and some do so hoping that they will make themselves seem more distinct because of the new mode of expression they have adopted. New modes seem especially good candidates for this sort of behavior since they offer such limited improvements over prior modalities that it makes many wonder what they are for if not for novelty. Some would say this of Twitter, for example, which allows users to send only short messages, albeit in a fashion that can be broadcast. In some views, tweets are like feeble blogs, the only advantage being that they can be updated anytime and anyplace via a mobile phone. Some of the technological artifacts of expression become a means for attaining distinction, too. The iPhone is perhaps the best current example of how keeping in touch by phone can allow someone to display status through the ownership of particular form of device. Owning one says something about the owner.

The history of communications channels certainly shows the claim being made again and again that new is better than the old and that the new will substitute what has come before. But the experience of using new channels leads people to discover that these new channels afford different opportunities for expression and for making different ties between themselves and those they communicate with. Twitter's success is not because it is a feeble version of a blog but because its limited length encourages a certain discursive brevity and because its instantness generates a feeling of connection. It might be short, but a tweet is a way of catching the breath of the blogosphere, even as it exhales.

Twitter might be something else too. It might be used more to create and sustain new forms of celebrity than to link friends. In the United Kingdom, for example, most tweets get read by no more than a handful of friends and fellow tweeters, whereas one or two get read by thousands. The people who create the latter Twitter feeds are either famous for other reasons (because they are TV personalities, for example) or for comments that have made them a Twitter celebrity in their own right, which draws new readers to their feeds. In those cases where celebrity is not at play, Twitter gets used not to mirror a network of real friends but to create a smaller network within these friendship sets. Here some members post more feeds than others and so become more honored within the friendship network than others as a consequence. One might say it is still a kind of celebrity, but it is a small-world kind (see Huberman et al.'s 2008 "Social Networks That Matter: Twitter under the Microscope").

Whether it is used to convey the blogosphere or celebrity, Twitter succeeds because it creates asymmetric ties. In this regard, it is distinct from the ties created by the essential mechanism of Twitter (short message service). The SMS (as a practice, not as a technological mechanism) has more symmetry ("I send to you, and you send one back"). That this is so attests also to how apparently similar modes of communication (texting and tweeting) afford different ways of binding.

THE METAPHYSICS OF THE SELF

The general point that the motivation behind using new channels may in large part be in an effort to make people distinct is, I think, not entirely untrue (although in saying this, I don't want to as pejorative as Baudrillard). People adopt new channels because doing so lets them deploy new tactics in an already enormously rich, complex, and delicate fabric of social ties. But it is precisely because this fabric is so delicate and refined that new modalities find their appeal. They allow people to create a new inflection in the performance of themselves.

Some commentators, such as Kenneth Gergen in his book, *Relational Being: Beyond Self and Community* (2009), suggest that this turn to expressive richness ought to presage a change in how we understand ourselves, our essential nature (see also Gergen 2000). In this view, an emphasis on performance in communication should lead us away from thinking that people have some kind of essence—a true inner self that they convey when communicating (like a cargo to be exchanged: "Give me who you are, and I will return the favor"). Rather, people are best thought of as produced through dialog. They are their communicative performance. This argument resonates through a whole range of thinkers who had no inkling of what our age of being in touch would enable, such as the Russian literary theorist Mikhail Bakhtin, who wrote mostly in the 1920s and 1930s. It was Bakhtin's fate to be internally exiled in the Soviet Union during a time when physical distance was meant to be a physical barrier to human connection and human dialogic performance, as he defined it.

But I am not convinced that the emergence of new expressive modalities does or ought to lead us to alter our sense of self. Although I admit the merits of the view and believe that one can exaggerate the fixity and cementlike form of the inner self at the expense of recognizing the performativity of the self, I also think a balanced view is required here. We need a view that can countenance both perspectives but without their extremes. Just as Baudrillard is too curmudgeonly when it comes to people seeking uniqueness (suggesting that their desire destroys anything genuine), so is Gergen too extreme in his view about what human identity might entail. It seems to me that we use new channels not because of a crude Baudrillardian desire to be unique but because we have a reasonable hope that we will be seen for what we are in our performance—as us, ourselves, and not mere consumers. Similarly, our delight and passion for new modes of expression and for new ways of dancing the dance of ourselves doesn't mean that we abandon any idea that we have a certain soul or essence that is us. We don't need a Gergen-like assertion of a new metaphysics of the self to recognize that we are unique as individuals and in terms of the web of social relations we create in our acts of communication. This makes the society of which we are a part just as it makes us, to some degree. No wonder that some explore new ways of creating expression and some seek new ways of leveraging old modes of communicating to demarcate who they are and who they are becoming.

As we try to avoid the excesses of some of these arguments, we should not forget that in trying to paint a portrait of themselves and give renewed vigor or subtlety to their relationships, people will find themselves in muddles, which can lead to confusions about identity and relationships. Sometimes new channels don't do what the user imagines, and sometimes they don't do what nonusers expect. As we saw in chapter 3, telepresence technologies didn't bring together people in a workplace to work as the inventors expected, but they could bring some people closer together romantically. But media space connectivity did not foster romance by itself. Romance flourished only when other vehicles (geographical proximity) allowed it to be cultivated. Because these two people were in the same lab, whether or not they were linked by digital connectivity, meant that they bumped into each other and shared the Dionysian aspects of the workplace.

CONCLUSION

I started this book with remarks about how many emails my colleagues and I receive each day. Too many, seemed to be my claim. We are all overloaded. But as the evidence and arguments that make up this book have shown, such an assertion is neither wholly accurate nor leading us toward the right way of thinking about the age of being in touch. Although I might be sent over a hundred emails a day, far fewer get through to my desktop because filters remove many. Our complaints about receiving too many messages at home (letters, phone messages, Facebook postings) were also somewhat misleading. We don't really get that many at home, and if we look at our overall use of time, we could invest even more effort into being in touch than we do already. But what we do at home when we keep our social networking sites up to date is not the same as when we deal with email and other forms of messaging at work. In both cases, we might use digital means to communicate, but what we are doing in each place, home and work, is better understood as different in purpose, in goal, and in the social fabric of which it is a part. It ought to be different too in how the benefits or otherwise are to be measured.

When we complain about email overload at work, we are sometimes complaining that there is too much work and that email is a distraction from that work. When we complain about our communications at home, we are complaining about other things—about the burdens of friendship or the monitoring of family affairs, for example. These are not the same. Moreover, when we engage in each of these different practices (work and family affairs), the *who* of us involved is different. At work, we are incumbents of an organizational role, and despite our occasional desire to disappear from view, we are visible to other work colleagues. They can message us requests for this or that piece of information or for advice on that action or this device because that is our responsibility. It's our job. But at home (or indeed anywhere other than work), the *who* of us in our acts of communication is different still. We can disown any responsibility for the housework if we can't be bothered, putting it off until tomorrow in a way that would cause frowns at work. One can disregard messages left on the home answering machine if one wants to ignore the world outside and put up one's feet. After all, when at home, one often puts effort into doing nothing—into the labor of rest (see Harper 2002, chapter 6). Nevertheless, one's children demand attention ("Play with me," "Help me with my homework," "Drive me to my club"). We have different labors at home. Who we are there is different than who we are at work.

This doesn't mean that there is only one home persona and one work persona. At home, one is at once a mother, a father, or a sibling, just as in the public domain, one is a friend or a member of a club. One's social identities are varied, just as are our private roles. Our use of communication technologies reflects that. But this use does not simply reflect this diversity as it is also constitutive of the selves in question. In some respects, we turn out to be our communication acts. We are, to some of our colleagues, little more than an email and to some of our relatives, little more than a recipient of invitations. But sometimes we are more than this. Our emails can create an inflection of ideas and hopes that alters the trajectory of our colleagues, even our companies. Our missives to the relatives honor them and the family of which we are all a part in a way that no physical presence could achieve. Sometimes families can come together only in the written word.

On any particular day, we might receive messages that seem to have no point, and we might complain that we don't seem to have enough time to service all our ways of being in touch. This may be just so, in some cases. But the purpose of this book has been to show how the texture of expression—the social ties that are created through communication acts, however mediated—is vast, delicate, and involved. It binds people through space and time into systems of mutually dependent acts and next turns. It can create vehicles to protect a sense of personal dignity and can allow that dignity to be mocked. It enables the portrayal of a sense of the past and a vision of a future. It ties people down and sometimes lets them feel free. The kinds of connections that people create between themselves have enormous range, color, and valence, and the use of new channels reflects and constitutes that diversity and gives it new intricacies.

To suggest that people communicate too much is like saying that people are bound to each other in too many ways and that the society that is thus produced is too awash with human connections. We might have opinions about what society we want to create and be a part of, and we need to recognize that making that society should not impose on us a burden that comes at too great a cost—forcing us to communicate with so many people that we cannot foster deep relationships with a handful of intimates or find space to honor the singularity of romance. We don't want to make our connections so febrile that we have no opportunity to show constancy and reliability, and we don't want our expressions to be so brief that we cannot convey elaborate reflections. We don't want to make a society out of Twitter feeds, after all. But we do want to make a society that is rich and diverse because we want to be rich in our human diversity. Just like the old fashioned written letter or a posting on a social network account, Twitter feeds do have various roles. It is not only society that we wish to make diverse and rich. It is us too. We don't need to buy into all the more extreme views about how people are really nothing but their communications to accept that what and how we communicate are essential elements of who we are. There may be other factors that drive our passions and form our characteristics, but our expressions and our expressive artfulness are certainly among the most important of the whole. In this sense, we fabricate ourselves through our use of various modes of being in touch.

We don't have only Twitter or a handful of mechanisms to construct a sense of ourselves and to convey ourselves to our web of friends, colleagues, and acquaintances. We have many. New mechanisms of being in touch extend the weave of being in touch, increasing as they do so the social fabric of our existence. That fabric is not simply geographic or temporal, but it binds us together to create a texture of human identity and connection. This book has shown that this texture has many forms, and the moral that derives from recognizing this texture is that our communication acts, in all their variety, make us all a part of the society of which we are both members and creators. That is why communication is the measure of our age. It is a measure of us and of what we do.

NOTES

1. Explanation of human communication seems peculiarly susceptible to conceptual muddles. As Michael Reddy noted some thirty years ago in his article "The Conduit Metaphor: A Case of Frame Conflict in Our Language about Language," it isn't just Locke who gets muddled. Many people do, including specialists when they cease doing communication acts and offer analyses of them.

2. See also E. B. Davies review of this in *Science in the Looking Glass: What Do Scientists Really Know?* (2003).

3. Sociologists are not alone in this focus, of course. Economists have been even more susceptible to this limited view. See, for example, Frances Cairncross's (1997) book *The Death of Distance: How the Communications Revolution Will Change Our Lives.*

4. Take, for example, "Friending," the practice of accepting a request for someone to join one's social network. This is a new act of communication, given a new delicacy to the webs of social connection. For discussion, see Charles Petersen, "In the World of Facebook."

5. One problem that we have when we seek such judgments about character is that the word is not the only one used to explore the who behind the acts. We use other words that have different properties that can make judgment difficult—or at least sometimes confusing. I am thinking here of the word *identity*, for example. Identity and character are not quite the same. Identity emphasizes an essence, and character emphasizes performance and action. When we are judging acts, we often evoke identity when we mean character, and this can lead to confusion. After all, we are more than what our passport says (the ultimate form of social identity), but a passport (a doing or a communication of a certain kind) might be all we have by which to judge a character.

6. The same can also happen with the use of mobile phones, which can allow teenagers to create walls that separate them from those they wish to hide from See Ito, Okabe, and Matsuda, *Personal, Portable, Pedestrian* (2005) and Ito's "Mobile Phones, Japanese Youth, and the Replacement of Social Contact" from 2003.

7. Although there are now more emails sent per capita than there were ever letters sent, some researchers have suggested that email is pushing up letter writing since the properties of email make the properties of the handwritten word all the more sensual and personal. See Harper, Palen, and Taylor 2005, *The Inside Text*, chapter 1.

REFERENCES

Baudrillard, J. 1970. *The Consumer Society: Myths and Structures.* Trans. C. Turner. London: Sage.

boyd, d. m. 2008. Taken out of context: American teen sociality in networked publics. PhD dissertation, University of California, Berkeley.

Cairncross, F. 1997. *The Death of Distance: How the Communications Revolution Will Change Our Lives.* Boston: Harvard Business School Press.

Churchill, E., and E. Goodman. 2008. (In)visible partners: People, algorithms, and business models on-line. In *Proceedings of EPIC 2008* (86–100). Arlington, VA: American Anthropological Association.

Davies, E. B. 2003. *Science in the Looking Glass: What Do Scientists Really Know?* Oxford: Oxford University Press.

Di Gennaro, C., and W. Dutton. 2007. Reconfiguring friendships: Social relationships and the Internet. *Information, Communication, and Society* 10(5): 591–618.

Gergen, K. 2000. *The Saturated Self: Dilemmas of Identity in Contemporary Life.* New York: Basic Books (originally published 1991).

Gergen, K. 2009. *Relational Being: Beyond Self and Community.* Oxford: Oxford University Press.

Habermas, J. 1991. *The Structural Transformation of the Public Sphere: An Inquiry into a Category of Bourgeois Society.* Cambridge, MA: MIT Press.

Habermas, J. 2006. Towards a United States of Europe. *Signandsight .com.* Available from http://sightandsound.com/features/676.htmel.

Harper, R., ed. 2002. *Inside the Smart Home.* Godalming: Springer.

Harper, R., L. Palen, and A. Taylor, eds. 2005. *The Inside Text: Social Perspectives on SMS.* Dordrecht: Kluwer.

Henkin, D. M. 2006. *The Postal Age: The Emergence of Modern Communications in Nineteenth-Century America*. Chicago: University of Chicago Press.

Huberman, B., D. Romero, and F. Wu. 2008. Social networks that matter: Twitter under the microscope. *First Monday* 14(1–5). Retrieved February 3, 2009, from http://firstmonday.org/htbin/cgiwrap/bin/ojs/index.php/fm/article/view/2317/2063.

Ito, M. 2003. Mobile phones, Japanese youth, and the replacement of social contact. Paper presented at the conference *Front Stage, Back Stage: Mobile Communication and the Renegotiation of the Public Sphere*, Grimstad, Norway, June 22–24.

Ito, M., D. Okabe, and M. Matsuda, eds. 2005. *Personal, Portable, Pedestrian: Mobile Phones in Japanese Life*. Cambridge, MA: MIT Press.

Lindley, S., R. Harper, and A. Sellen. 2009. Desiring to be in touch in a changing communications landscape: Attitudes of older adults. In *SIGCHI Conference on Human Factors in Computing Systems (CHI 2009)* (1693–1702). New York: ACM.

Meyrowitz, J. 1985. *No Sense of Place: The Impact of Electronic Media on Social Behavior*. Oxford: Oxford University Press.

Nass, C., and S. Brave. 2005. *Wired for Speech: How Voice Activates and Advances the Human-Computer Relationship*. Cambridge, MA: MIT Press.

Nielson NetRatings. 2005. Available online at http://www.nielsonnetratings.com/pr/pr_051013_uk.pdf.

Penrose, R. 1989. *The Emperor's New Mind: Concerning Computers, Minds, and the Laws of Physics*. Oxford: Oxford University Press.

Penrose, R. 1994. *Shadows of the Mind: A Search for the Missing Science of Consciousness*. Oxford: Oxford University Press.

Peters, J. D. 1999. *Speaking into the Air: A History of the Idea of Communication*. Chicago: University of Chicago Press.

Petersen, C. 2010. In the world of Facebook. *New York Review of Books* LVII (3):8–11.

Reddy, M. 1979. The conduit metaphor: A case of frame conflict in our language about language. In A. Ortony, ed., *Metaphor and Thought* (284–325). Cambridge: Cambridge University Press.

Rettberg, J. W. 2008. *Blogging*. Cambridge: Polity Press.

Sennett, R. 1986. *The Fall of Public Man*. London: Penguin (originally published 1977).

Sennett, R. 1998. *The Corrosion of Character*. New York: Norton.

Shirky, C. 2008. *Here Comes Everybody: The Power of Organising without Communications*. London: Allen Lane.

Standage, T. 1998. *The Victorian Internet*. New York: Berkeley Books.

INDEX

Aakhus, Mark, 119
Absent presence, 120
Aggregators, 27–28
Agre, P. E., 104n8
Algorithms, 69, 83, 98, 104n9, 156
America Calling: A Social History of the Telephone to 1940 (Fisher), 14
Ames, M., 98
Anderson, Ben, 14
Anthropology, 84, 119, 123, 131, 135, 144, 174, 195, 224n4
Assumptions, 221–223
Austin, John L., 10, 198–199
Ayer, A. J., 186, 193

Badge systems, 86–90
Bailey, George, 46
Bakhtin, Mikhail, 262
Banta, Marta, 52–53
Baudrillard, Jean, 7, 259, 261–262
Bayesian prediction, 79
Bedroom culture, 42
Behaviorism, 78, 83, 175
 cybernetics and, 94–98, 188n4, 240
 dualism and, 77, 79, 84, 174–175
 friendship and, 176–177 (*see also* Friendship)
 Glancephones and, 159–169, 172–173, 176–184, 193–194, 206, 211, 243
 mischief and, 64, 166, 173, 218–219
 moral requirement to express and, 207–211
 networking and, 252–255 (*see also* Social networking)
 perceived benefits of communication and, 234–238
 saying too much and, 215–221
 sociality and, 211–214
 voluntaristic theory of action and, 175

Behaviorism (cont.)
Whereabouts clocks and, 159, 168–173, 176–182, 193–194, 204, 206, 211, 243
Benkler, Y., 46–47
Berg, S., 130
Berking, Helmuth, 130–131, 137
Bertschi, Stefan, 119
Bible, the, 18
Big Brotherism, 63
BillG, 10
Bittner, Egon, 70–71, 220
BlackBerrys, 3, 11
Blackburn, Simon, 198
Blogging (Rettberg), 257
Blogs, 3
aggregators and, 27–28
art and, 24
blogosphere and, 25–27, 29, 93, 113, 257, 260–261
class and, 110
communications paradox and, 12–13, 20–29, 37, 42, 46–50, 53
content and, 156, 167, 181
corpspeak and, 24–25
diaries and, 24, 26
as digital commons, 146
emotion and, 146
filtering and, 93
identity and, 25, 257–258
impact of, 13
initial excitement over, 13
intimacy and, 23–24, 28
invention and, 76, 93, 97
Kierkegaardian irony of, 27
measurement issues and, 28–29
mobile technology and, 110, 113, 144–146
organizational communication and, 23–25
particular properties of, 255–256
person-to-person communication and, 42
reality and, 48
really simple syndication (RSS) feeds and, 26–28
responses to, 258
Sennett on, 256–257
social structure and, 110
spontaneity and, 26
texture and, 242, 248, 255–261
topics of, 25–28
Blog: Understanding the Information Reformation That Is Changing Your World (Hewitt), 13
Blown to Bits: How the New Economics of Information Transforms Strategy (Evans and Wurster), 46
Bolter, Jay David, 155
Books, 11
Bourdieu, Pierre, 174–175, 180, 194–195, 199, 245
Bowling Alone (Putnam), 111, 144
boyd, danah, 37, 249–250

Brave, Scott, 236
Broadcasting, 17, 23, 25, 28, 36–37, 49, 260
Brown, Barry, 119
Busyness, 4
  communications paradox and, 11, 29–30, 35, 40–42
  efficiency and, 1
  historical perspective on, 28–35, 41–42
  knowledge workers and, 42–43, 51
  status and, 35–36
  time measurement studies and, 28–33

Call forwarding, 176, 242
Capitalism, 111
Cartesian dualism, 77–79
Cartesian Linguistics (Chomsky), 186
Castells, Manuel, 43, 45–47, 49, 112, 238
Cavanagh, Christine, 11–12, 42
Cavell, Stanley, 200
"Changing Cultures of Written Communication" (Höflich and Gebhardt), 141
Changing Times: Work and Leisure in Postindustrial Society (Gershuny), 14, 35
Character construction, 168, 172–174, 178–179
Choice, 180–182
Chomsky, Noam, 186
Christianity, 18
Churchill, E., 250
City in Your Pocket, Birth of the Mobile Information Society (Kopomaa), 122
Clarissa (Richardson), 18
Clark, Andy, 109
Click-throughs, 93
Clifford, James, 224n4
Coaxial cables, 63, 69
Cognitivists, 186–187
Communication
  absent others and, 114
  act of, 28
  beneficial, 11–12, 234–238
  blogs and, 3, 12–13 (*see also* Blogs)
  body mechanics and, 47, 53–54, 60–67, 71–80, 83–86, 94–101, 109, 158–160, 169, 174–176, 222, 239, 243–245
  books and, 11
  choice and, 180–182
  community and, 110–113, 117–120, 124–125, 132–137, 140–143, 147, 188n7
  connectivity and, 51, 76, 80, 117–118, 121, 123, 193, 241, 263
  context and, 114, 118, 129, 157–158, 172, 178, 182, 216, 219–220, 249–251, 256
  couriers and, 59
  cultural theory and, 14, 54
  cybernetics and, 94–98, 188n4, 240

Communication (cont.)
desire for, 5
disembodied, 47
emotion and, 46, 132, 139–145, 144, 256
empirical fallacies and, 234–241
etiquette and, 160–161, 166
excitement over new channels of, 6, 12, 180, 183, 233, 246, 258–263, 266
eye contact and, 67–71
face-to-face, 67–71, 110, 137, 139, 160, 162, 217, 237
folk wisdom and, 195
forms of, 9
gesture and, 72, 74, 101
greetings sequence and, 160–162, 165–166, 212, 251
historical perspective on, 13–20
home environment and, 91–92
identity and, 19 (*see also* Identity)
as indulgence, 158
innovation in, 12–13
instant messaging (IM) and, 1, 3, 11, 20, 28, 28–29, 34, 37–38, 65, 74, 80, 83, 197, 224, 231, 237, 239, 242, 248
insults and, 160
interaction model for, 60–67
Internet and, 11, 14, 23, 33–38, 45–49, 55nn2, 59, 111, 113, 116, 122, 138, 142–145, 238–243, 252, 255, 259
isolation and, 111, 122
linguistics and, 184–187, 193
love and, 50, 98, 130, 132, 134, 141, 144, 181, 206, 208, 210–211, 218
meaning of word, 9–10
measurement issues and, 13–20
mobile technology and, 132–135 (*see also* Mobile technology)
motivations for, 157–160
narrative and, 14–18, 43, 52, 154, 165, 173–174, 258
newspapers and, 11, 43, 49–51
new ways of, 3–5
obsessiveness over, 146–147
older people and, 39, 229–233, 254
organizational, 23–25
paradoxes of, 22, 28, 43–45, 70, 81, 100, 102, 184, 253
as passing information, 9
person-to-person, 42, 110, 118, 121–123
postal culture and, 18–20
private space and, 4, 19, 23, 41, 46, 60, 87, 101–102, 119, 125, 146, 177, 214, 252–253, 256–257, 265
prolixity and, 180, 217
prosaic doings and, 7, 15, 36, 64, 80, 204–208
radio and, 43, 49
real-world approach to, 194–195, 204
relationships and, 20–23

rudeness and, 160, 176, 210, 251
saying too much and, 215–221 (*see also* Overload)
Skype and, 97–98
sociality and, 41, 49, 65, 113, 211–214, 220, 251, 257
social networking and, 20, 23, 37, 39, 42–43 (*see also* Social networking)
spatial flows of, 112–113
status and, 35–36
synonyms and, 14, 42, 59, 98, 110, 182, 185
tactile experiences and, 3
teenagers and, 21, 33, 35–42, 65, 123, 126–131, 134–137, 170–172, 249–254, 268n6
telecommunication term and, 59
television and, 36, 39–40, 49–51, 165, 177, 252
texture and, 196–197 (*see also* Texture)
time for, 28–35
touch and, 26–27 (*see also* Touch)
triage systems and, 92–93
uneven distribution of, 35–37
as vehicle for intentions, 202
virtual address books and, 126–127, 162, 174, 241–242
virtue and, 15, 20–23, 48, 146, 235
vision of human user and, 76–80
work setting and, 92

Communication Power (Castells), 43, 112
Communications in the 21st Century (Nyiri), 119
Communications theory, 184
Community, 119–120, 188n7
absent others and, 114
agricultural context and, 117
altered geography of, 111
bowling alone and, 111
change from local, 110–113
connectivity and, 51, 76, 80, 117–118, 121, 123, 193, 241, 263
etiquette and, 160–161, 166
face-to-face communication and, 110
friendship rituals and, 127–130
gifting and, 113, 130–132, 136, 141–142, 158, 166, 180, 187, 242
global, 110–113
information flows and, 111
macro/micro view of, 115
mobile technology and, 117–118 (*see also* Mobile technology)
organic-to-mechanical shift in, 111
place-based connectivity and, 117–118
ritual and, 132–135
sense of belonging and, 117
space/time decoupling and, 113–114

Community (cont.)
technological impacts on, 111
Wellman and, 47, 113–118, 121–127, 142–143, 241
Computers
algorithms and, 155–156
as calculators, 153
computer-human landscape and, 155–157
desktop publishing and, 153
extended functions of, 153–154
graphical icons and, 156
hackers and, 154–155
hypertextuality and, 155
imposition of mind upon, 154
mainframe, 153
operating system (OS) and, 155–156
personal, 20, 59, 88, 92, 103n1, 153–156
programming and, 153–157
TCP-IP stack and, 156
Turkle on, 154–157, 188n1
wearable, 153
Computer-supported collaborative work (CSCW), 64, 105n16
Conlon, Tom, 154
Connections: New Ways of Working in the Networked Organization (Sproull and Kiesler), 12
Connectivity, 76, 80, 193
communications paradox and, 51
cultural effects of, 122–124
role-to-role, 116–118, 123–124
texture and, 241, 263
widening of, 121–122
Consumerism, 258–259
Consumer Society, The: Myths and Structures (Baudrillard), 7, 259
Content
avoiding silence and, 185
cognitivists and, 186–187
Diamond and, 201–204
digital cameras and, 153
glancing and, 68, 70, 74, 81, 91, 109, 158–169, 179
language and, 184–187, 198–203
programming and, 153–156
science and, 193–195, 201–202
skepticism and, 199–202
starting point for, 198–203
telementation fallacy and, 184–185
wiling away time and, 185
Context
content and, 157–158, 172, 178, 182
mobile technology and, 114, 118, 129
networking and, 251
philosophy and, 216, 219–220
texture and, 249–251, 256
Corrosion of Character, The (Sennett), 256
Couriers, 59
Course in General Linguistics (de Saussure), 186
Cowardice, 137–140

C-slate, 101
"C-Slate: Exploring Remote Collaboration on Horizontal Multi-touch Surfaces" (Izadi), 72
Cultural theory, 14, 17–18, 54–55
Cybernetics, 94–98, 188n4, 240
Cybernetics: or the Control and Communication in the Animal and the Machine (Wiener), 94
Cyborgs, 109

Data capture devices, 87–89
De Sola Pool, I., 50
Derrida, Jacques, 9–10
de Saussure, Ferdinand, 186
Determinism, 243
Diamond, Cora, 20–22, 201–204
Diaries, 24, 26, 40
"Difficulty of Reality and the Difficulty of Philosophy, The" (Diamond), 201
di Gennaro, Corinna, 225n5, 242–243
Digital cameras, 153
Dilman, I., 186
Distinction: A Social Critique of the Judgment of Taste (Bourdieu), 180
Dourish, Paul, 198
Dualism, 76–80, 84, 174–175
Durkheim, 111
Dutton, William, 36, 225n5, 242
Dystopia, 7

E-commerce, 36
Economics of Attention, The: Style and Substance in the Age of Information (Lanham), 93–94
Economy, 180
Editor window, 63
Efficiency, 1, 4, 6, 85, 97, 166, 180, 216
Email, 213–214
- beneficial, 12
- BlackBerrys and, 3, 11
- changed attitude over, 12
- communications paradox and, 34–38, 42–44
- contacting old friends and, 3–4
- content and, 156, 178–181, 189
- daily volume of at Microsoft, 1, 10–11, 34
- filters and, 2, 10–11, 34, 264
- increasing use of, 10
- initial excitement over, 12
- innovation of, 12–13
- interruptions and, 1, 43
- invention and, 85, 93, 98
- junk, 10–13
- measurement and, 28–29
- overload and, 1, 10, 12–13, 215–216, 264–266
- person-to-person communication and, 42
- saying too much and, 215–221
- status and, 35–36
- texture and, 237–238, 248, 257, 263, 268
- triage systems and, 92–93

Emotion, 46, 256
blogs and, 146
etiquette and, 160–161, 166
glancing and, 162
insults and, 160
laughter and, 64, 166–167, 173, 183, 194
mobile technology and, 132, 139–145, 144
rudeness and, 160, 176–177, 210, 251
Erbring, Lutz, 47–48
Ergonomics, 95
Essay Concerning Human Understanding, An (Locke), 184
Estaunie, Edouard, 59
Ethnography of Malinowski, The (Young), 135
Etiquette
emotion and, 160–161, 166
Glancephones and, 163, 166
greetings sequence and, 160–162, 165–166, 212, 251
rudeness and, 176–177
Evans, Philip, 46
Exchange servers, 11
Expression, 195–196

Facebook, 3, 207
communications paradox and, 23, 37–38, 42
content and, 157
mobile technology and, 146
texture and, 249–254, 264, 268
Face-to-face communication, 137, 237
community and, 110
content and, 160
eye contact and, 67–71
greetings sequence and, 162
saying too much and, 217
taking turns and, 139
video conferencing and, 67–71
Fall of the Public Man, The (Sennett), 256
Familiar Letters on Important Occasions (Richardson), 18
Family, 98, 207, 211
communications paradox and, 39
content and, 169–173, 176–181
extended, 112–113
mobile technology and, 110–113, 122, 146
nuclear, 112–113
texture and, 244, 264–265
Whereabouts clocks and, 170–173, 177
Feminism and Youth Culture (McRobbie), 42
Feudal systems, 111
Filters
blogs and, 93
email and, 2, 10–11, 34, 264
as fire fighting, 2
quality control and, 85, 90, 93
triage systems and, 92–93
Finland, 122
Fisher, Claude, 14

Flânerie, 120–121
Flickr, 37, 253–254
Fodor, Jerry, 84
Forms of life, 204
For Sociology: Renewal and Critique in Sociology Today (Gouldner), 134–135
Fortunati, Leopoldina, 120
France, 250
Friending, 268n4
Friendship
 behaviorism and, 176–177
 geography and, 249
 media space and, 176–177
 memories and, 132–133
 mindfulness and, 176
 mobile technology and, 127–130
 networking and, 206–207, 249 (*see also* Social networking)
 sociality and, 211–214
Friendship routines, 127–130

Game play, 229
Garfinkel, Harold, 204–205
Gates, Bill, 10
Gaver, William, 63
Gebhart, Julian, 119, 141
Gemeinschaft, 111, 116, 188n7
Geography of touch, 73–74
Gergen, Kenneth, 7, 120, 122, 261–263
Germany, 121, 250
Gerontology publications, 229
Gershuny, Jonathan, 14, 35–36
Gesselschaft, 111, 116
Gestures, 72, 74, 101
Gifting, 113, 242
 choice and, 180–181
 content and, 158, 166, 180, 187
 mobile technology and, 130–132, 136, 141–142
 SMS and, 166
Gladstone, William, 17–18
Glancephones, 193–194, 206, 211, 243
 attention getting and, 165
 caller's mood and, 161
 as camera phones, 163
 choice and, 181–182
 construction of character and, 178–179
 content and, 159–169, 172–173, 176–184
 etiquette and, 160–161, 163, 166
 greetings sequence and, 160–161
 idea of fitting and, 160–164
 interruptions and, 164–165
 research findings on, 164–168
 ring tones and, 161–162
 sensory depth and, 159–160
 software applications for, 163–164
 value from, 172
Glancing, 109
 attention getting and, 165
 construction of character and, 179
 content and, 158–169, 179

Glancing (cont.)
emotion and, 162
Glancephones and, 159–169, 172–173, 176–184, 193–194, 206, 211, 243
greetings sequence and, 160–162, 165–166, 212, 251
interruption assessment and, 163
invention and, 68, 70, 74, 81, 91
Glotz, Peter, 119
Goodman, E., 250
Goodwin, C., 165
Google, 74–75, 96, 103n1, 169
Gouldner, Alvin W., 134–135
GPS systems, 91
Green, Nicola, 119
Greetings sequence, 160–162, 165–166, 212, 251
Grice, Paul, 199, 217, 219
Grint, Keith, 104n4
Groupware, 64
Gulia, M., 143

Habermas, Jurgen, 257
Habitus
content and, 175–182
philosophy and, 193, 199, 211, 219
texture and, 245, 254
Hacker, P. M. S., 79, 195
Hackers, 154–155
Hamill, Lynne, 33–34, 65, 119, 121, 123
Hanfling, Oswald, 223n2
Harper, R., 65, 119, 121, 123, 130, 230, 265, 268n7
Harris, Roy, 184–185, 194–195
Harrison, Steve, 103n2, 104n5
Hartmann, Maren, 119
Haythornthwaite, Caroline, 47
Helpser, Ellen J., 36
Henkin, David, 18–20, 22, 31, 44, 110, 159, 230, 236
Here Comes Everybody: The Power of Organizing without Communications (Shirky), 48, 146, 157
Hewitt, H., 13
Hillygus, Sunshine, 47–48
Hodges, S., 87
Höflich, Joachim, 119, 141
Horvitz, Eric, 90–91, 94
How Blogs Are Changing the Way Businesses Talk with Customers (Scoble and Israel), 24
How to Do Things with Words (Austin), 10, 199
Huberman, B., 261
Human Computer Interaction (HCI), 95, 197–198
Human expression
badge systems and, 86–90
behaviorism and, 78, 83–84, 175, 188 (*see also* Behaviorism)
blogs and, 3, 12–13, 20, 22–29 (*see also* Blogs)
body mechanics and, 47, 53–54, 61–62, 71–80, 83–86,

94–101, 109, 159–160, 169, 174–178, 222, 239, 243–245
body/mind issues and, 76–80, 86
choice and, 180–182
community and, 110–113, 117–120, 124–125, 132–137, 140–143, 147, 188n7
connectivity and, 51, 76, 80, 117–118, 121, 123, 193, 241, 263
construction of character and, 168, 172–174, 178–179
cowardice and, 137–140
cultural theory and, 14, 54
cybernetics and, 94–98, 188n4, 240
cyborgs and, 109
data capture devices and, 87–89, 91
diaries and, 24, 26, 40
dualism and, 76–80, 84, 174–175
embarrassment and, 113
embodied action and, 174–175
emotion and, 162 (*see also* Emotion)
empirical fallacies and, 234–241
essence of self and, 7
etiquette and, 160–161, 166
family and, 39, 98, 110–113, 122, 146, 169–173, 176–181, 207, 211, 244, 264–265
fragility of social relations and, 135–137
friendship rituals and, 127–130
future and, 6–7
gesture and, 72, 74, 101
gifting and, 113, 130–132, 136, 141–142, 158, 166, 180, 187, 242
glancing and, 68, 70, 74, 81, 91, 109, 158–169, 179
greetings sequence and, 160–162, 165–166, 212, 251
habitus and, 175–182, 193, 199, 211, 219, 245, 254
historical perspective on, 13–20
as information processing, 49–55
innovation and, 5, 30, 80, 95, 141, 238
insults and, 160
interaction model for, 60–67
Internet and, 11, 14, 23, 33–38, 45–49, 55nn2,3, 59, 111, 113, 116, 122, 138, 142–145, 238–243, 252, 255, 259
isolation and, 111, 122
laughter and, 64, 166–167, 173, 183, 194
letters and, 2, 12–24, 28–35, 38, 41–43, 50 (*see also* Letters)
love and, 50, 98, 130, 132, 134, 141, 144, 181, 206, 208, 210–211, 218
mobile technology and, 110 (*see also* Mobile technology)
moral requirement to express and, 207–211

Human expression (cont.)
motivations for, 157–160
narrative and, 14–18, 43, 52, 154, 165, 173–174, 258
older people and, 39, 229–233, 254
pointings and, 72, 74, 158
pragmatism and, 61–62, 79, 82–84, 86, 90, 206, 240
processing limits of humans and, 61, 85–93
programming and, 154–157
prosaic doings and, 7, 15, 36, 64, 80, 205–208
qualia and, 88–90
reading speed and, 50
ritual and, 132–135
rudeness and, 160, 176, 210, 251
sociality and, 41, 49, 65, 113, 220, 251, 257
social networking and, 252–261 (*see also* Social networking)
sociology of, 241–243
touch and, 45, 59, 113, 199, 223, 260 (*see also* Touch)
Turing model and, 77–79, 82–85, 95–98, 104n9, 110, 174, 184, 231–232
video and, 37, 54, 62–63, 67–71, 74, 81, 87, 97–98, 102, 116, 164, 189, 197
vision of human user and, 76–83
Wave and, 74–75

Human Nature: The Categorical Framework (Hacker), 195
Hume, David, 200
Huurdeman, Anton, 59
Hypertext markup language (HTML), 156
Hypertextuality, 155

Identity, 7, 92
blogs and, 257–258
communications paradox and, 19, 25
community and, 110–118 (*see also* Community)
connectivity and, 51, 76, 80, 117–118, 121, 123, 193, 241, 263
content and, 162, 172, 177–182, 188
expanded functions of technology and, 154
loss of, 113
metaphysics of self and, 261–263
mobile technology and, 113, 120, 123, 146
networked, 146
self-esteem and, 43–44
texture and, 233, 245–246, 257, 262–267, 268n5
Turkle on, 154–157, 188n1
Incongruency of perspective, 67–68
Information and Communication Technologies in Society: E-living in a Digital Europe (Anderson), 14

Information flows, 39
Innovation, 5, 80, 95, 238
communications paradox and, 12–13, 30
email and, 12–13
initial excitement over, 12–13
mobile technology and, 141 (*see also* Mobile technology)
Inside Text, The: Social, Cultural, and Design Perspectives on SMS (Harper), 119
Instant messaging (IM), 1, 3, 80, 83
communications paradox and, 11, 20, 28–29, 34, 37–38
philosophical view and, 197, 224
shared whiteboards and, 63–67
social networking and, 63–65
texture and, 231, 237, 239, 242, 248
ubiquitous nature of, 230–231
Wave and, 74–75
Insults, 160
Intelligence
behavior source and, 78
qualia and, 88–90
Turing model and, 77–79
Intelligent algorithms, 69–70
Interactional geography, 73
Internet
altered geography and, 111
changes from, 14
communications paradox and, 11, 14, 23, 33–38, 45–49, 55nn2,3
computer code and, 155–156
consumption data on, 36–37
displaced television watching and, 36
effort used on, 143–144
lessening of social contact and, 122
measurement issues and, 14
mobile technology and, 111, 113, 116, 122, 138, 142–145
Skype and, 97–98
teenage use of, 36–37
texture and, 238–243, 252, 255, 259
UK use of, 35, 36
URLs and, 155
use of term, 59
Wellman and, 142–143
World Wide Web and, 2, 4, 20, 24–27, 37, 51–52, 60, 75, 92, 94, 155–157, 163
Internet and Everyday Life, The (Wellman and Haythornthwaite), 47
Internet Galaxy, The: Reflections on the Internet, Business, and Society (Castells), 45, 238
Interruptions
constant, 1, 61, 85, 91–92, 162–166
email and, 1, 43
glancing and, 163
home environment and, 91–92
letters and, 43
Microsoft and, 1–2
work setting and, 92

Intimacy, 48, 214, 221
blogs and, 23–24, 28
defining acts for, 47–48
empirical fallacies and, 234–241
etiquette and, 160–161, 166
good-night messages and, 127
letters and, 16–23, 41–42, 230, 233
overload and, 266
perceived benefits of communication and, 234–238
ritual and, 127
social structure and, 110
texting and, 126, 130, 132, 136, 141
video connections and, 81
Invention, 1, 5–6, 8, 61, 92, 99–100, 103
badge systems and, 86–90
communications paradox and, 13, 19, 28, 31, 52
consumer relationship and, 66
content and, 156, 159, 167, 188
data capture devices and, 87–89, 91
EuroPARC and, 60, 62–65, 67–75, 81–82, 86–89, 103n2
initial excitement over, 12–13
innovation and, 5, 30, 80, 95, 141, 238
interaction and, 62, 66–67
justification of, 66–67, 89, 100
knowledge work and, 63–64
media space and, 63
mobile technology and, 109, 115
more-is-better approach and, 60
philosophy and, 194–196, 206, 222–223
pragmatism and, 83–84
shared whiteboards and, 63–65
texture and, 238–241, 244–245, 263
Turing model and, 77–79, 82–85, 95–98
video and, 37, 54, 62–63, 67–76, 81, 87, 97–98, 102, 116, 164, 189, 197
vision of human user and, 76–83
visual record system, 86–87
iPhone, 260
Isolation, 111, 122
Israel, Shel, 24–25, 27, 48
Ito, Mizuko, 119, 268n6
Izadi, Shahram, 72

Japan, 119, 123
Jefferson, G., 160
Johnson, Samuel, 17
Judgment, 7–8
assumed shared knowledge and, 218
of character, 221
communications paradox and, 24, 28, 44
content and, 163, 165, 176, 179–180, 184, 187
economy of expression and, 219
information source and, 219

invention and, 78–80, 104n8
mobile technology and, 144–145
networking and, 253 (*see also* Social networking)
older people and, 232
philosophy and, 196, 201–203, 221–223
saying too much and, 215–221
texture and, 232, 244–248, 255, 268n5
Junk email, 10–13

Katz, James, 55n2, 118–119
Kavanaugh, Andrea L., 47, 55n3
Keys, 70
Kierkegaard, 27
Kiesler, Sara, 12
Knowledge work
communications paradox and, 42–43, 51, 63–64
management tools and, 116
video conferencing and, 67–74, 67–76
Kopomaa, Timo, 122
Kraut, Robert, 55n3

LamdaMOO, 157
Landow, Goerge P., 155
Language, 255
Austin and, 10, 198–199
blogs and, 25
cognitivists and, 186–187
communications theory and, 184
content and, 184–187, 198–203
Diamond and, 201–204
empirical fallacies and, 234–241
formal sterilities of, 47
Harris and, 186
inventive landscapes and, 102
myths of, 231
natural, 237
philosophy and, 198, 204, 223nn3,4
programming and, 153–156
prosaic doings and, 7, 15, 36, 64, 80, 205–208
subjectivism and, 200
synonyms and, 14, 42, 59, 98, 110, 182, 185
Turing model and, 78, 231–232
Language Myth, The (Harris), 184
Language, Truth, and Logic (Ayer), 186
Lanham, Richard, 93–94
Lasen, Amparo, 119, 121
Laughter, 64, 166–167, 173, 183, 194
Lectures on Conversation (Sacks), 205
Lenhart, Amanda, 36
Letters, 96, 102, 137
American society and, 18–20
as art form, 16–17, 29, 41
beneficial, 12
British society and, 15–18
broadening genres of, 15–20
changed attitude over, 12
communications paradox and, 12–24, 28–35, 38, 41–43, 50

Letters (cont.)
content and, 181
cultural practice and, 17–18
diligence needed for, 230
English literature and, 15–16
epistles and, 18
form and, 16
frequency of, 31–33
as grub street publications, 16–17
historical perspective on, 13–20, 28–35, 235–236
honor and, 233
identity and, 19
interruptions and, 43
intimacy and, 16–23, 41–42, 230, 233
love notes and, 50, 141
machinery of life and, 16
measurement issues and, 28–33
moral implication of, 2
narrative and, 16
New Testament, 18
Penny Black stamp and, 30–31
personal, 19–20
person-to-person communication and, 42
philosophy and, 197, 213
physical separation and, 61
postal culture and, 18–20
private space and, 19–20
published, 16–17
real time and, 75
reflection and, 230
relationships and, 20–23, 41–42
Royal Mail and, 30–31
as self-expression, 20
sense of place and, 15–16
social ladder and, 16
style and, 16
texts and, 43
texture and, 230, 233–237, 240, 259, 264–268
time for, 28–35
travelogues and, 16
virtue and, 15, 20–23, 235
volume and, 31–33
Lévi-Strauss, C., 84
Lindley, S., 230
Ling, Rich, 119, 121, 123
Linguistics, 184–187, 193
LinkedIn, 254
Lissak, Robin, 46
Locke, Chris, 119
Locke, John, 184, 193, 198
London, 64
Love, 98, 181
letters and, 50, 141
philosophy and, 206, 208, 210–211, 218
use of mobile technology and, 130, 132, 134, 144
Lutz, Catherine, 144

Madden, Mary, 36
Maffesoli, Michel, 65, 69
Magee, B., 186
Magic in the Air: Mobile Communications and the Transformation of Social Life (Katz), 119

"Making Love in the Network Closet: The Benefits and Work of Family Videochat" (Ames), 98
Malcolm, Norman, 90, 223n2
Managing Your Email: Thinking Outside the Box (Cavanagh), 11, 42
Marcus, George, 224n4
Marr, David, 79
Marvin, Carolyn, 116
Marx, Karl, 111
Matsuda, Misa, 119, 268n6
Mauss, Marcel, 130–131
McRobbie, Angela, 42
Me++ (Mitchell), 169
Measurement studies
aggregation issues and, 38–39
blogs and, 28–29
context and, 14–15
email and, 28–29
failure to recognize change and, 14
historical perspective and, 13–20
letters and, 28–33
rush to apply new metrics and, 248–249
television and, 51
time and, 28–33, 37–40
Media space
data capture devices and, 87–91
friendship and, 176–177
incongruency of perspective and, 67–68
narrative and, 174
networking and, 252–255 (*see also* Social networking)
qualia and, 88–90
shared whiteboards and, 63–67
video conferencing and, 67–71
visual record system and, 86–87
Whereabouts clocks and, 169–170
Memory, 223n2, 247
friendships and, 132–133
invention and, 86–90
qualia and, 88–90
Memory and Mind (Malcolm), 90
Meyrowitz, Joshua, 252
Microsoft, 25, 156, 238
Bill Gates and, 10
daily volume of email at, 1, 10–11, 34
interruption issues in, 1–2
invention and, 60, 75, 104
Milinowski, Bronislaw, 135
Mischief, 64, 166, 173, 218–219
Mitchell, William, 169
Mobile Communication in Everyday Life: An Ethnographics View (Höflich and Hartmann), 119
Mobile Communication: Perspective and Current Research Fields (Höflich and Gebhardt), 119
Mobile Communications: Re-negotiation of the Social Sphere (Ling and Pederson), 119
Mobile Connection, The (Ling), 123

Mobile society, 43
Mobile technology, 3, 59, 103n1
absent others and, 114
absent presence and, 120
bodily pattern and, 174–178
caller's mood and, 161
choice and, 180–182
closer social sense from, 126–127
connectivity and, 51, 76, 80, 117–118, 121, 123, 193, 241, 263
construction of character and, 168, 172–174, 178–179
contact evidence for, 121–124
content and, 153, 160–162, 166–168, 176, 179
cowardice and, 137–140
as digital mediation, 139–140
effort used on, 143–144
embodied action and, 174–175
emotion and, 132, 139–145, 144
etiquette and, 160–161, 166
fragility of social relations and, 135–137
framing mechanism for, 142–143
friendship routines and, 127–130
gifting and, 130–132, 136, 141–142
Glancephones and, 159–169, 172–173, 176–184, 193–194, 206, 211, 243
greetings sequence and, 160–162, 165–166, 212, 251
as invigorating tool for social relations, 122
iPhone and, 260
liberation of, 118
literature on, 118–121
as new stage, 120
paradoxical delights and, 111–115, 118–127, 137–140, 143–145, 150
pull of fashion and, 120–121, 124–125
rethinking social impact of, 124–127
ring tones and, 121, 161–162
ritual and, 132–135
role-to-role connectivity and, 116–118, 123–124
rudeness and, 176–177
space/time decoupling and, 113–114
spatiality and, 119–120
teenagers and, 123, 126–131, 134–137
tethering effects and, 122–124
text messages and, 1, 42 (*see also* Text messages)
texture and, 230, 241–242, 260, 268
virtual address books and, 126–127, 162, 174, 241–242
Whereabouts clocks and, 159, 168–170, 173, 176–182, 193–194, 204, 206, 211, 243

Mobile World: Past, Present and Future (Hamill and Lasen), 119
Mobilities (Urry), 43, 112
"Models of Attention in Computing and Communication: From Principles to Applications" (Horvitz), 90–91
Moral issues, 5
communications paradox and, 9, 18, 24, 27
content and, 166–167, 172–175, 180, 187, 189
implicativeness and, 221–222
invention and, 61–62, 70–71, 78, 81–82, 99
judgment and, 221–223 (*see also* Judgment)
mobile technology and, 136–137, 142, 145
moral requirement to express and, 207–211
older people and, 232
perceived benefits of communication, 234–238
philosophy and, 193–195, 199, 212, 217, 220–221
singular acts and, 221–222
texture and, 231–235, 241, 244, 258, 267
"Movement-Space: The Changing Domains of Thinking Resulting from the Development of New Kinds of Spatial Awareness" (Thrift), 112
Multimedia messaging service (MMS), 164
"Must We Say What We Mean?" (Bittner), 71
Mutual referent, 68
Myth of the Paperless Office (Sellen), 50

Nagel, K., 94, 104n13
Narrative, 52, 258
the communicating human and, 45–49
content and, 154, 165, 173
defining, 43
historical perspective on, 14–18
letters and, 16
media space and, 174
mobile society and, 43
network society and, 43
theory of narrative form and, 174
travelogues and, 16
Narrowcasting, 23, 28
Nass, Clifford, 236
Natural-Born Cyborgs: Minds, Technologies and the Future of Human (Clark), 109
Neighborhoods and, 117–118
Networks
digitally mediated sociality and, 248–261 (*see also* Social networking)
individualism and, 116
technological affordances and, 116

Network society, 43
Neuman, W. Russell, 49–52
New Delhi, 64
Newman, M., 87
Newspapers, 11, 43, 49–51
New Tech, New Ties (Ling), 119
New Testament, 18
New York, 64
Nie, Norman, 47–48
Nielson NetRatings, 250
Nietzshe, Friedrich, 194
No Sense of Place: The Impact of Electronic Media on Social Behavior (Meyrowitz), 252
Nyiri, Kristóf, 119

Okabe, Daisuke, 119, 268n6
Older people, 39, 229–233, 254
"On Computable Numbers, with an Application to the Entscheindungsproblem" (Turing), 77–78
Operating system (OS), 155–156
Ottoman clock, 44
Outline of a Theory of Practice (Bourdieu), 174
Outlook, 10–11
Overeducation, 36
Overload
 balance of things and, 1–2
 communicative burden and, 167
 constant interruptions and, 1, 61, 85, 91–92, 162–166
 defining, 11–12
 email and, 1, 10, 12–13, 215–216, 264–266
 empirical fallacies and, 234–241
 informated environment and, 215
 information processing and, 52
 intimacy and, 266
 measurement issues and, 13–20
 metaphysics of, 93–99
 networked individualism and, 116
 older people and, 232
 perception of, 41
 processing limits of humans and, 61, 85–93
 really simple syndication (RSS) feeds and, 1–2, 26–27
 research focus on, 2–8
 saying too much and, 215–221
 tension from, 5
 text messages and, 1, 42, 126–128, 141, 257
 threshold for, 2–3
 time measurement and, 28–33
 voice messages and, 2
 volume and, 1, 4
Overwork, 36

Parson, Talcott, 84, 175, 188n4
Paul, Apostle, 18
Pederson, Per, 119
Penny Black stamp, 30–31
Penrose, Roger, 240
Pentland, Alex, 223n3
Perpetual Contact: Mobile Communication, Private Talk, Public Performance (Katz and Aakhus), 119

Personal computers (PCs)
communications paradox and, 20
content and, 153–156
invention and, 59, 88, 92, 103
mobile technology and, 113
Personal, Portable, Pedestrian: Mobile Phones in Japanese Life (Ito, Okabe, and Matsuda), 119, 268n6
Person-to-person communication, 42 (*see also* Communication; Friendship communications paradox and)
greetings sequence and, 160–162, 165–166, 212, 251
mobile technology and, 110, 118, 121–123 (*see also* Mobile technology)
Peters, John Durham, 53–54, 188n7, 238
Phaedrus (Plato), 198
Philosophical Investigations (Wittgenstein), 84, 204
Philosophy, 4
behaviorism and, 78, 83–84, 175, 188
cognitivists and, 186–187
communications paradox and, 9–10, 27, 53
community and, 110–113
consumerism and, 258–259
content and, 186–187, 193–204, 217, 223n2
determinism and, 243
dualism and, 77, 79, 84, 174–175
embodied action and, 174–175
expression and, 195–196
goodness of ambition and, 193–194
habitus and, 175–182, 193, 199, 211, 219, 245, 254
invention and, 79, 83–84, 90
language and, 198, 204, 223n3, 224n4
metaphysics of self and, 261–263
mind/body dualism and, 174
moral issues and, 193–195, 199, 212, 217, 220–221
pragmatism and, 61–62, 79, 82–84, 86, 90, 206, 240
prosaic doings and, 7, 15, 36, 64, 80, 205–208
qualia and, 88–90
reductionism and, 109, 175
science and, 78–79, 84, 94, 98, 184, 186, 194–195, 201–202, 240
sensibility and, 196
skepticism and, 199–202
structuralism and, 84
subjectivism and, 200
texture and, 196–198, 234, 239–240, 258–259
view of human user and, 76–80
Physical geography, 73

Piccinini, G., 78
Pitch of Philosophy, A: Autobiographical Exercises (Cavell), 200
Plant, Sadie, 120
Plato, 194, 198
Poetry, 9
Pointings, 72, 74, 158
Postal Age, The: The Emergence of Modern Communications in Nineteenth-Century America (Henkin), 18–19, 236
Postal culture, 18–20
Pragmatism, 61–62, 79, 82–84, 86, 90, 206, 240
Private space
  blogs and, 23
  communications paradox and, 19, 23, 41, 46
  content and, 177, 214
  invention and, 60, 87, 101–102
  letters and, 19–20
  mobile technology and, 119, 125, 146
  social networking and, 253
  texture and, 252–253, 256–257, 265
  time for reflection in, 4
Programming, 153–156
Prolixity, 180, 217
Prosaic doings, 7, 15, 36, 64, 80, 205–208
Pull of fashion, 124–125
Puro, Jukka-Pekka, 119–120
Putnam, Robert, 111, 144

Qualia, 88–90
Quine, W. V. O., 186, 193

Radio, 43, 49
Rankin, Alexandra, 36
Reading, 50–51
Really simple syndication (RSS) feeds, 1
  aggregators and, 27–28
  blogs and, 26–28
  changed Web site content and, 2
  threshold for, 2–3
"Reconfiguring Friendships: Social Relationships and the Internet" (di Gennaro and Dutton), 242–243
Reddy, Michael, 267n1
Reductionism, 109, 175
Relational Being: Beyond Self and Community (Gergen), 7, 261–262
Relationships. *See* Intimacy
Religion, 18, 250
Rettberg, Jill W., 257
Rheingold, Howard, 112
Rice, Ronald, 55n2
Richardson, Samuel, 18
Ring tones, 161–162
Rise of the Network Society, The (Castells), 43
Ritual, 132–135
Rodzilla, J., 13
Royal Mail, 30–31
Rudeness, 160, 176–177, 210, 251

Sacks, Harvey, 160, 205
Schegloff, E., 160
Schmidt, Kjeld, 105n16
Schwalbe, Will, 11, 42
Scoble, Robert, 24–25, 48
Searle, John, 199
Second Self, The: Computers and the Human Spirit (Turkle), 154
Second Self, The: Computers and the Human Spirit, Twentieth Anniversary Edition (Turkle), 156–157
Sellen, Abi, 50, 230
Send: The How, Why, When and When Not of Email (Schipley and Schwalbe), 11, 42
Sennett, Richard, 256
Shanker, S., 78
Shannon, Claude, 184
Shared whiteboards, 63–67
Sheutz, M., 79
Shipley, David, 11, 42
Shirky, Clay, 48–49, 51, 146, 157, 249, 253
Short message service (SMS), 119, 141, 166, 181, 242, 261
"Signature Event Context" (Derrida), 9–10
Skepticism, 199–202
Skinner, B. F., 83–84
Skype, 97
Sleep, 34, 36, 38, 44
Smart Mobs: The Next Social Revolution (Rheingold), 112
Smart mob technology, 112
Smith, Aaron, 36
Sociality
  communications paradox and, 41, 49
  digitally mediated, 248–261
  invention and, 65
  mobile technology and, 113
  philosophy and, 211–214, 220
  scale of, 250–252
  texture and, 251, 257
Social networking, 98
  blogs and, 255 (*see also* Blogs)
  communications paradox and, 20, 23, 37, 39, 42–43
  consumerism and, 258–259
  content and, 157, 179, 183
  danger and, 253–254
  digital commons and, 146
  digital scale of, 250–252
  instant messaging (IM) and, 63–65
  mobile technology and, 116, 118, 143–146
  person-to-person communication and, 42
  philosophy and, 206–207, 213, 220, 225
  photos and, 253
  public, 252–255
  superimpositioning of, 45–46
  texture and, 242, 249–255, 261, 264, 266–267
  time spent on, 249
"Social Networks that Matter: Twitter under the Microscope" (Huberman), 261

Social structure, 110
Social System, The (Parsons), 84
Sociology
absent others and, 114
American society and, 18–20
anthropology and, 84, 119, 123, 131, 135, 144, 174, 195, 224n4
bedroom culture and, 42
behaviorism and, 78, 83–84, 175, 188
British society and, 15–18
communications paradox and, 14–15, 43, 46, 49, 51
community and, 110–113, 117–120, 124–125, 132–137, 140–143, 147, 188n7
construction of character and, 168, 172–174, 178–179
content and, 159, 166, 173–174, 177, 183, 188n7
cowardice and, 137–140
cultural practice and, 17–18
cultural theory and, 14, 54
etiquette and, 160–166, 176–177, 212, 251
of expression, 241–243 (*see also* Human expression)
family and, 39, 98, 110–113, 122, 146, 169–173, 176–181, 207, 211, 244, 264–265
fragility of social relations and, 135–137
fragmentation and, 113
friendship routines and, 127–130
gifting and, 113, 130–132, 136, 141–142, 158, 166, 180, 187, 242
habitus and, 175–182, 193, 199, 211, 219, 245, 254
historical perspective and, 13–20
identity and, 7, 19, 25, 92 (*see also* Identity)
individualized experience and, 112–113
invention and, 65, 84, 104n4
measurement issues and, 13–20, 28–33
mobile technology and, 110–116, 122–125, 129–136, 140–144, 147
moral issues and, 5, 9, 18, 24, 27 (*see also* Moral issues)
narrative and, 14–18, 43, 52, 154, 165, 173–174, 258
new codes of conduct and, 112
philosophy and, 193, 204–208, 224n4
postal culture and, 18–20
pull of fashion and, 120–121, 124–125
ritual and, 132–135
solidarity and, 113
structure and, 110
teenagers and, 21, 33, 35–42, 65, 123, 126–131, 134–137, 170–172, 249–254, 268n6
television and, 39–40
texture and, 229, 233, 241–245, 256, 268n3

touch and, 26–27 (*see also* Touch)
transcending time and space with, 20–21
tribalism and, 46, 65, 69
virtue and, 20–23
Sociology of Giving (Berking), 130–131
Spatial Formations (Thrift), 111–112
Speaking into the Air: A History of the Idea of Communication (Peters), 53, 188n7, 238
Speech Acts (Searle), 199
Speech-act theory, 10
Spreading the Word: Groundings in the Philosophy of Language (Blackburn), 198
Sproull, Lee, 12
Standage, Tom, 259
Start Problem-Solving with Prolog (Conlon), 154
"Strategy and the New Economics of Information" (Evans and Wurster), 46
Structuralism, 84
Structural Transformation of the Public Sphere, The (Habermas), 257
Studies in Ethnomethodology (Garfinkel), 204–205
Studies in Social Interaction (Sudnow), 70
Studies in the Way of Words (Grice), 199, 217
Subjectivism, 200
Sudnow, David, 70, 162
Synonyms, 14, 42, 59, 98, 110, 182, 185

Tactile experience, 3
"Taken out of Context: American Teen Sociality in Networked Publics" (boyd), 37
Taylor, Alex, 51, 130
Taylored Lives: Narrative Productions in the Age of Taylor, Veblen, and Ford (Banta), 52–53
TCP-IP communications, 156
Technological affordances, 116
Technology
altered experience from, 8, 111
beneficial, 11–12
calculation and, 153
choice and, 180–182
computers and, 155–156 (*see also* Computers)
connectivity and, 51, 76, 80, 117–118, 121, 123, 193, 241, 263
consumerism and, 258–259
cybernetics and, 94–98, 188n4, 240
enabling, 4–5
essence of self and, 7
excitement over new channels of communication and, 6, 12, 180, 183, 233, 246, 258–263, 266
explosion of communication and, 46

Technology (cont.)
future and, 6–7
Glancephones and, 159–169, 172–173, 176–184, 193–194, 206, 211, 243
GPS and, 91
initial excitement over, 12–13
innovation and, 5, 30, 80, 95, 141, 238
invention and, 1, 5–6, 8, 13 (*see also* Invention)
justification of, 66
loss of community from, 111
mobile, 110 (*see also* Mobile technology)
older people and, 39, 229–233, 254
programming and, 153–156
pull of fashion and, 120–121, 124–125
quicker communications and, 5–6
Skype and, 97–98
smart mob, 112
telecommunications and, 59, 111, 116
Turing model and, 77–79
Turkle on, 154–157, 188n1
as umbilical cord, 59
values and, 7, 12, 15 (*see also* Values)
vision of human user and, 76–83
Whereabouts clocks and, 159, 168–170, 173, 176–182, 193–194, 204, 206, 211, 243
Teenagers, 65
bedroom culture and, 42
communications paradox and, 21, 33, 35–42
content and, 170–172
friendship routines and, 127–130
Internet use by, 36–37
location of parents and, 170–173
mobile technology and, 123, 126–131, 134–137
social networking and, 42, 250–252 (*see also* Social networking)
text messages and, 127–128
texture and, 249–254, 268n6
ubiquitous messaging of, 230–231
Whereabouts clocks and, 170–173
Teens and Social Media (Lenhart, Madden, Rankin, and Smith), 36
Telecommunication term, 59
Telementation fallacy, 184–185
Telephones, 237
cost of long-distance calls and, 15
historical perspective on, 13–15
mobile, 121 (*see also* Mobile technology)
modems and, 156
Television
behavioral effects of, 252–253
communications paradox and, 25, 33, 36, 39–40, 43, 49–51

consumption data on, 36–37
content and, 165, 177
digital, 122
interactive, 122
Internet watching and, 36
lessening of social contact and, 122
measurement studies of, 51
monopolizing influence of, 39–40
texture and, 237, 252, 260
time spent on, 39–40
Text messages, 1, 42, 257
cowardice and, 137–140
doing friendship and, 129
emotion and, 132, 139–145
fragility of social relations and, 135–137
gifting and, 130–132, 141–142
good-night messages and, 127
intimacy and, 126, 130, 132, 136, 141
ritual and, 132–135
teenagers and, 127–128
Texture
assumptions and, 221–223
basic tenets of, 232
body mechanics and, 47, 53–54, 61–62, 71–80, 83–86, 94–101, 109, 159–160, 169, 174–176, 222, 239, 243–245
content and, 198–203
deepened sense of, 197
Diamond and, 201–204
forms of life and, 204
human-computer interaction (HCI) literature and, 197–198
judgment and, 202, 221–223
moral requirement to express and, 207–211
saying too much and, 215–221
science and, 193–195, 201–202
seams and, 197–198, 221
skepticism and, 199–202
sociality and, 211–215, 220, 251, 257
strengths of, 196–197
words as part of world and, 203–207
Thousand Tribes, A: How Technology Unites People in Great Companies (Lissak and Bailey), 46
Thrift, Nigel, 111–112
Thumb Culture: The Meaning of Mobile Phones for Society (Glotz and Bertschi), 119
Thus Spoke Zarathustra (Nietzsche), 194
Time
consumption patterns in, 37–40
historical perspective on, 28–35
home environment and, 91–92
Internet use and, 33
knowledge workers and, 42–43, 51
lack of, 37–40
leisure, 33–34
letters and, 28–35
loss of, 52

Time (cont.)
measurement studies and, 28–33, 37–40
networking and, 249
Ottoman clock and, 44
reading and, 50–51
sleep and, 34, 36, 38, 44
television viewing and, 39–40
work and, 33–34
Time of the Tribes, The: The Decline of Individualism in Mass Society (Maffesoli), 65
Tönnies, 111
Touch
being in, 27, 43, 54, 60, 74, 76, 96, 99, 110, 112, 115, 142, 147, 157, 182, 207, 222, 230, 234–235, 246, 259, 262–267
being out of, 92, 147, 229
communications paradox and, 27, 43–47, 54
conceptual synonyms for, 110
content and, 157–159, 168, 182
geography of, 73–74
intimacy and, 110 (*see also* Intimacy)
invention and, 59–60, 65, 70–76, 88, 91–92, 96, 99, 102
keeping in, 4, 45, 59, 65, 113, 194, 199, 223, 234, 260
mobile technology and, 110–115, 142, 147
older people and, 39, 229–233, 254
philosophical analysis of, 194, 199, 206–207, 217, 222–223
pointings and, 72, 74, 158
social structure and, 110
texture and, 229–230, 234–235, 246, 254, 259–267
"Tracking the Flow of Information" (Pool), 50
Tracking the Flow of Information into the Home (Neuman), 49
Transportation, 111
Travelogues, 16
Triage systems, 92–93
Tribalism, 46, 65, 69
Turing, Alan
approach to understanding intelligence by, 77–79
behaviorism and, 78, 83–84
Cartesian dualism and, 79
computer algorithms and, 83
cybernetics and, 95–97
decision problem and, 77–78
influence of, 79
media space and, 174
pragmatism and, 83–84
vision of human user and, 77–79, 82–85, 95–98, 104n9, 110, 174, 184, 231–232
Turkle, Sherry, 154–157, 188n1
Turow, Joseph, 47, 55n3
Twitter, 3, 12, 93, 144–145, 167, 259–261, 266–267

Uniform resource locators (URLs), 155

United Kingdom, 15–18
- Internet use in, 35–36
- mobile technology and, 121
- television viewing and, 39–40
- texture and, 250, 260
- time consumption patterns in, 37–40

United States
- Civil War and, 18
- consumption patterns in, 49
- Gold Rush and, 18
- identity and, 19
- Internet use in, 36–37
- letters and, 22
- postal culture of, 18–20

University of Toronto, 116
Urry, John, 43, 111–112

Values, 7
- communications paradox and, 12, 15, 27–29, 42, 44
- content and, 157, 169–172, 175, 178, 183, 186–187
- invention and, 78, 80, 82, 96
- mobile technology and, 121–122, 130, 133, 141
- performative, 244–245
- philosophy and, 192, 202, 212, 214
- sensitivity and, 245–246
- texture and, 231–237, 244–246, 256, 259

Victorian Internet, The (Standage), 259
Victorians, 235, 259
Video, 197
- communications paradox and, 37, 54
- content and, 116, 164, 189
- invention and, 62–63, 67–71, 74, 81, 87, 97–98, 102
- media space and, 63
- streaming, 69
- visual record system and, 86–87

Video conferencing
- eye contact issues and, 67–71
- following movement and, 69–70
- geography of touch and, 73–74
- gesture and, 72
- incongruency of perspective and, 67–68
- intelligent algorithms and, 69–70
- interactional geography and, 73
- keys and, 70
- knowledge work and, 67–76
- mutual referent and, 68
- object-recognition software and, 72
- physical geography and, 73
- user concept and, 72–73

Video-mediated communication (VMC), 62–63
Video tunnels, 68
Vincent, J., 121
Virtual address books, 126–127, 162, 174, 241–242
Virtue
- intimacy and, 48
- letters and, 15, 20–23, 235
- networked identity and, 146

Vision: A Computational Investigation into the Human Representation and Processing of Visual Information (Marr), 79
Voice messages, 2
Volume, 1, 4
as amount of words, 98
communications paradox and, 10–13, 31, 34–35, 43, 45, 49
content and, 166–167, 182
email and, 35
invention and, 60, 62, 98, 101–102
letters and, 31–33
measurement issues and, 13
mobile technology and, 109
philosophical view and, 211, 222–223
texture and, 232, 235–236
UK Internet use and, 35

Walpole Horace, 16–18
Wave, 74–75, 96
Wealth of Networks, The (Benkler), 46
Wellman, Barry
absent others and, 114
community and, 47, 113–118, 121–127, 142–143, 241
Internet studies and, 142–143
little boxes and, 116
networked individualism and, 116
role-to-role connectivity and, 116–118, 123–124
space/time decoupling and, 113–114
technological affordances and, 116
We've Got Blog: How Blogs Are Changing Our Culture (Rodzilla), 13
When Old Technologies Were New: Thinking about Electronic Communications in the Nineteenth Century (Marvin), 116
Whereabouts clocks
choice and, 181–182
construction of character and, 179
content and, 159, 168–170, 173, 176–182
family and, 170–173, 177
Harry Potter book series and, 168
location information and, 168–170, 173, 176–182, 193–194, 204, 206, 211, 243
media space and, 169–170
philosophical view of, 193–194, 204, 206, 211
research findings on, 170–173
teenagers and, 170–173
texture and, 243
time zones and, 168–169
value from, 171–173

Where the Action Is: The Foundations of Embodied Interaction (Dourish), 198
White, Geoffrey, 144
Wiener, Norbert, 94–95, 97, 110
Windows 95, 156
Wired for Speech: How Voice Activates and Advances the Human-Computer Relationship (Nass and Brave), 236
Wired Homestead, The (Turow and Kavanaugh), 47
Wireless World: Interdisciplinary Perspectives on the Mobile Age (Harper, Brown, and Green), 119
Wittgenstein, Ludwig, 84, 204–205
Wooffitt, R., 188n2
Woolgar, Steve, 104n4
Word and Object (Quine), 186
Work time, 33–34, 36
World War I era, 135
World War II era, 36
World Wide Web, 2, 4
  communications paradox and, 20, 24–27, 37, 51–52
  content and, 155–157, 163
  invention and, 60, 75, 92, 94
Wurster, Thomas S., 46

Xerox EuroPARC
  data capture devices and, 88–89
  invention and, 60, 62–65, 67–75, 81–82, 86–89, 103n2
  memory problems and, 88
  visual record system of, 86–87
Xerox PARC, 60

Young, Michael, 135
YouTube, 3, 37

Zuboff, S., 215